Organic Strawberry Production Manual

Technical Editors

Steven T. Koike, Carolee T. Bull, Mark Bolda, Oleg Daugovish

2012

University *of* California Agriculture and Natural Resources

Funding for writing this publication provided in part by the California Department of Food and Agriculture *Buy California Initiative,* the California Department of Pesticide Regulation, and the University of California Sustainable Agriculture Research and Education Program.

cdfa
CALIFORNIA DEPARTMENT OF FOOD AND AGRICULTURE

To order ANR publications and other products, visit the ANR Communication Services online catalog at http://anrcatalog.ucdavis.edu or phone 1 (800) 994-8849. You can also place orders by mail or FAX or request a printed catalog of our products from

University of California
Agriculture and Natural Resources
Communication Services
1301 S. 46th Street
Richmond, California 94804-4600

Telephone 1 (800) 994-8849
(510) 655-2195
FAX (510) 665-3427
E-mail: anrcatalog@ucdavis.edu

Publication 3531
ISBN-13: 978-1-60107-750-9
Library of Congress Control Number: 2012933812

Cover photos: Kathy Keatley Garvey (front, back top), Surendra Dara (back bottom).

UC PEER REVIEWED This publication has been anonymously peer reviewed for technical accuracy by University of California scientists and other qualified professionals. This review process was managed by the ANR Associate Editor for Pest Management—Agriculture, James J. Stapleton.

Printed in the United States of America on recycled paper.

3m-pr-5/12-WJC/RW

WARNING ON THE USE OF CHEMICALS

Pesticides are poisonous. Always read and carefully follow all precautions and safety recommendations given on the container label. Store all chemicals in their original labeled containers in a locked cabinet or shed, away from foods or feeds, and out of the reach of children, unauthorized persons, pets, and livestock.

Recommendations are based on the best information currently available, and treatments based on them should not leave residues exceeding the tolerance established for any particular chemical. Confine chemicals to the area being treated. THE GROWER IS LEGALLY RESPONSIBLE for residues on the grower's crops as well as for problems caused by drift from the grower's property to other properties or crops.

Consult your county agricultural commissioner for correct methods of disposing of leftover spray materials and empty containers. Never burn pesticide containers.

PHYTOTOXICITY: Certain chemicals may cause plant injury if used at the wrong stage of plant development or when temperatures are too high. Injury may also result from excessive amounts or the wrong formulation or from mixing incompatible materials. Inert ingredients, such as wetters, spreaders, emulsifiers, diluents, and solvents, can cause plant injury. Since formulations are often changed by manufacturers, it is possible that plant injury may occur, even though no injury was noted in previous seasons.

MEASUREMENT CONVERSION

U.S. Customary	Conversion factor for U.S. Customary to Metric	Conversion factor for Metric to U.S. Customary	Metric
LENGTH			
inch (in)	2.54	0.394	centimeter (cm)
foot (ft)	0.3048	3.28	meter (m)
AREA			
acre (ac)	0.4047	2.47	hectare (ha)
VOLUME			
gallon (gal)	3.785	0.26	liter (l)
MASS			
ounce (oz)	28.35	0.035	gram (g)
pound (lb)	0.454	2.205	kilogram (kg)
ton (T)	0.907	1.1	metric ton (t)
pound per acre (lb/ac)	1.12	0.89	kilogram per hectare (kg/ha)
ton per acre (T/ac)	2.24	0.446	metric ton per hectare (t/ha)
TEMPERATURE			
Fahrenheit (°F)	°C = (°F – 32) ÷ 1.8	°F = (°C x 1.8) + 32	Celsius (°C)

Contents

Preface

The commercial production, harvesting, and marketing of strawberries are complex undertakings that necessitate the consideration of many factors and the handling of many challenges. Organic strawberry production is further complicated by the use of alternative methods, the need to forego the use of conventional tools and inputs, and the overall orientation toward the use of biologically based systems. In addition, both conventional and organic systems must provide fruit that meets high standards of quality and is competitively priced in local, regional, and national markets.

The purpose of this manual is to bring together available information and research in support of the production, disease and pest management, postharvest handling, and marketing of organic strawberries in California and the certification of organic strawberry farms. We hope that the manual will give information and insights to a diverse audience, including growers, pest control advisers, consultants, other industry professionals, marketers, researchers with federal, state, and university institutions, individuals who are involved with the certification of organic farms, and other interested persons.

This *Organic Strawberry Production Manual* should be used in combination with other University of California references on strawberries, which include *Integrated Pest Management for Strawberries, Second Edition* (UC ANR Publication 3351) and *UC IPM Pest Management Guidelines for Strawberry* (UC ANR Publication 3468). Like any biological system, the strawberry production system is very complex, and our understanding of its intricacies is far from complete. Some parts of this manual therefore also draw attention to areas and subjects that would benefit from additional study and research.

When using this manual, we must remind you to follow all labels and instructions regarding the use of all farm products and inputs. Check with your local agricultural commissioner regarding the use of pest control materials. Consult closely with your organic certifier and certifying agency regarding all production decisions and inputs. In no way do we intend that this manual be used apart from the certification program in effect in your area.

— Steven T. Koike, Carolee T. Bull,
Mark Bolda, and Oleg Daugovish

Authors and Editors

Technical Editors

Steven T. Koike, UC Cooperative Extension Plant Pathology Farm Advisor, Monterey County

Carolee T. Bull, USDA Research Plant Pathologist, USDA ARS Crop Improvement and Protection Research Unit, Salinas, California

Mark Bolda, UC Cooperative Extension Strawberry and Caneberry Farm Advisor, Santa Cruz County

Oleg Daugovish, UC Cooperative Extension Strawberry and Vegetable Crop Farm Advisor, Ventura County

Short Course Coordinators and Initial Technical Editors

Sean L. Swezey, Lecturer, UC Santa Cruz Department of Environmental Studies

Janet C. "Jenny" Broome, formerly Associate Director, UC Sustainable Agriculture Research and Education Program; currently Senior Research Plant Pathologist, Driscoll Strawberry Associates

Jeri Ohmart, Food Systems Program Assistant, UC Sustainable Agriculture Research and Education Program

David Chaney, DEC Education Services, Corvallis, Oregon; formerly with UC Sustainable Agriculture Research and Education Program

Contributing Authors

Jianlong Bi, UC Cooperative Extension Entomology Farm Advisor, Monterey County

Vanessa Bogenholm, formerly Pest Control Adviser, VB Agricultural Services, Watsonville

Mark Bolda, UC Cooperative Extension Strawberry and Caneberry Farm Advisor, Santa Cruz County

Janet C. "Jenny" Broome, formerly Associate Director, UC Sustainable Agriculture Research and Education Program; currently Senior Research Plant Pathologist, Driscoll Strawberry Associates

Carolee T. Bull, USDA Research Plant Pathologist, USDA ARS Crop Improvement and Protection Research Unit, Salinas, California

David Chaney, DEC Education Services, Corvallis, Oregon; formerly with UC Sustainable Agriculture Research and Education Program

Surendra Dara, UC Cooperative Extension Strawberry and Vegetable Crops Farm Advisor, San Luis Obispo and Santa Barbara Counties

Richard De Moura, Extension Research Associate, UC Davis Department of Agricultural and Resource Economics

Steven A. Fennimore, Extension Specialist, UC Davis Department of Plant Sciences

Mark Gaskell, UC Cooperative Extension Small Farms and Specialty Crops Farm Advisor, San Luis Obispo and Santa Barbara Counties

Stephen R. Gliessman, Professor, UC Santa Cruz Department of Environmental Studies

Ray Green, formerly Organic Program Manager, California Department of Food and Agriculture

Karen Klonsky, Extension Specialist, UC Davis Department of Agricultural and Resource Economics

Steven T. Koike, UC Cooperative Extension Plant Pathology Farm Advisor, Monterey County

Frank N. Martin, Research Plant Pathologist, USDA ARS Crop Improvement and Protection Research Unit, Salinas, California

Elizabeth Mitcham, Director, Postharvest Technology Center, Extension Specialist and Pomologist, Department of Plant Sciences, UC Davis

Daniel C. Mountjoy, USDA Natural Resources Conservation Service Area Resource Conservationist, Salinas, California

Joji Muramoto, Associate Researcher, UC Santa Cruz Department of Environmental Studies

Richard Smith, UC Cooperative Extension Vegetable Crops and Weed Science Farm Advisor, Monterey County

Krishna V. Subbarao, Extension Specialist/Plant Pathologist, UC Davis Department of Plant Pathology

Sean L. Swezey, Lecturer, UC Santa Cruz Department of Environmental Studies

L. Ann Thrupp, Manager of Sustainability and Organic Development, Fetzer Winery, Mendocino, California

Laura Tourte, UC Cooperative Extension Farm Management Advisor/County Director, Santa Cruz County

Acknowledgments

Strawberries are among the most important fruit crops in California, and organically produced and consumed strawberries have been steadily increasing in popularity and prominence. This manual on the organic production and marketing of strawberries is part of an effort to support and increase this trend. The manual is also a model of cooperation and collaboration between diverse agencies, industries, researchers, growers, and many others. In 1998, research teams received a Sustainable Agriculture Research and Education Program (SAREP) grant and started a Biologically Integrated Farming Systems (BIFS) initiative intended to investigate, develop, and demonstrate methods for organic strawberry production. In 1999, a Biological Agricultural Systems in Strawberry–Organic Agricultural Systems in Strawberry (BASIS-OASIS) Pest Management Innovators work group was formed to research and identify biological alternatives for use in strawberry cropping systems and pest control. A manual on organic strawberry production was initially conceived by this BASIS-OASIS group. The authors acknowledge Bill Liebhardt and Jenny Broome (former SAREP administrators) for their support of the BIFS project and Carolee T. Bull (Project Manager) for leading the BASIS-OASIS work group.

These efforts and other activities came together and provided information and background for courses in organic strawberry production. In Monterey County in 2003, the University of California, USDA, and industry professionals presented a UC-sponsored short course and field tour on organic strawberry production. A similar course was subsequently offered in Ventura County in 2004. We acknowledge Sean Swezey for organizing these courses. We acknowledge the co-sponsors of these courses: UC Cooperative Extension in Santa Cruz County (Laura Tourte, Director), Monterey County (Sonya Hammond, Director), and Ventura County (Larry Yee, former Director); UC Davis Extension (Debbie Roberts and Jim Lapsley); the California Department of Food and Agriculture (*Buy California Initiative* program); the California Strawberry Commission; Ecological Farming Association; SAREP (David Chaney and Jeri Ohmart); and the UC Agriculture and Natural Resources (UC ANR) Organic Farming Research Group.

Many members of the strawberry industry have contributed valuable insights, experience, and assistance to the writing of this manual. We acknowledge and thank Tom Am Rhein, Vanessa Bogenholm, Jim Cochran, Larry Eddings, Rod Koda, Laurence London, Clint Miller, Miles Reiter, and the California Strawberry Commission.

This manual represents the collective experience and knowledge of more than 20 researchers and strawberry experts. Contributors include industry professionals and university academics, campus-based researchers and county-based extension researchers, representatives of federal and state agencies, and individuals experienced in economics, farm management, pest management, strawberry production, and general agriculture.

Finally, the manual authors thank the University reviewers and the UC ANR Communications Services and Information Technology team for assistance in publishing this work.

— Steven T. Koike, Carolee T. Bull, Mark Bolda, and Oleg Daugovish

ECONOMIC PERFORMANCE AND MARKETING CONSIDERATIONS

PHOTO: KATHY KEATLEY GARVEY

1 Growth of the Organic Strawberry Industry in California

KAREN KLONSKY

To really understand the size and growth of the organic strawberry industry in California, you need to see it in context with national production and sales. We will start out by looking at the size and growth of organic food sales in the United States, with an emphasis on the importance of organic fruit, and identifying sales, growth, and market outlets. Then we will go on to characterize the size and growth of organic production at the farm level in California, again with emphasis on the importance of fresh produce. Finally, we will look at the actual numbers of growers and acres and the location of organic strawberry production operations in California.

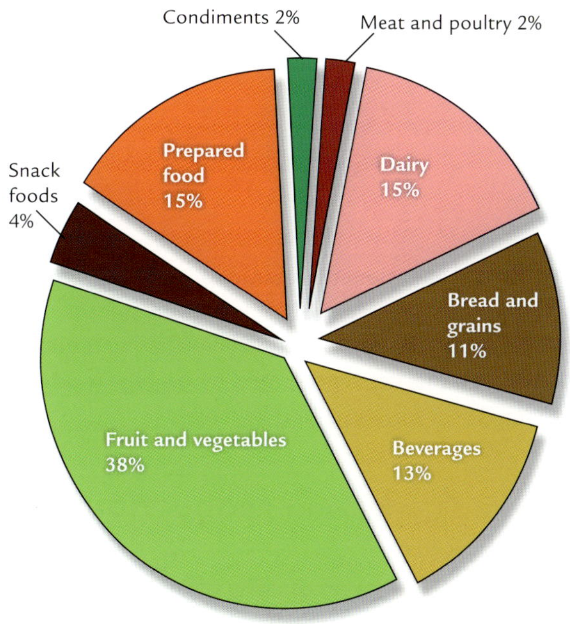

Figure 1.1. Organic food category shares, from $24.8 billion total sales in 2009. *Source:* Organic Trade Association.

ORGANIC FOOD SALES IN THE UNITED STATES

Retail sales of organic food in the United States exceeded $24.8 billion in 2009 (figure 1.1) (Nutrition Business Journal 2010). Growth over the past 10 years (2001 to 2010) has been steady at an impressive 15 to 20 percent a year. But it is critical to keep in mind that while organic food sales increased by 130 percent from 2000 to 2005 and by 75 percent between 2005 and 2009, organic food still accounted for only 3.7 percent of total food sales in 2009. This is an impressive increase, though, from its 1.2 percent share in 2000.

Looking at U.S. marketing outlets in 2009, less than half of the organic food sold went through natural or health food stores (40 percent), slightly more than half went through mass-market outlets (54 percent), and the remainder (6 percent) went through farmers markets or other direct sale channels (Nutrition Business Journal 2010). The share of organic sales posted by various distribution channels has changed dramatically over the past 10 years. From 1991 to 1995, natural and health food stores represented over two-thirds of sales and farmers markets and other direct sales accounted for another fourth of sales. Conventional retailers handled only 7 percent of all organic sales (Dimitri and Greene 2002).

The largest category of organic food sales over time has been fruits and vegetables (Nutrition Business Journal 2010). In 2009, this category accounted for 38 percent of all organic food sales, at $9.5 billion (figures 1.1 and 1.2). Sales of fruits and vegetables grew by 11.4 percent between 2008 and 2009, for an increase of $970 million in sales. Looking more closely at fruit and vegetable sales for the period, 92 percent was sold into the fresh market, frozen produce made up another 4 percent, and the remaining sales were of canned (3 percent) and dried (1 percent) produce (Nutrition Business Journal 2010).

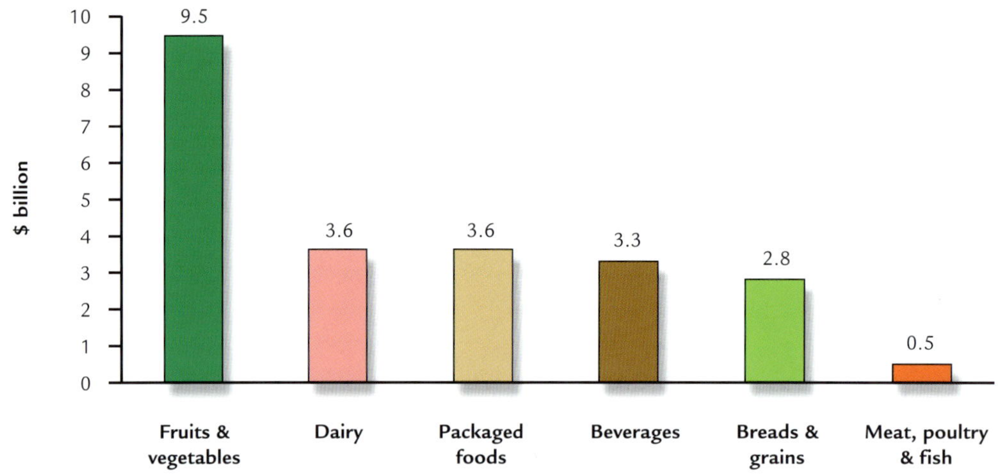

Figure 1.2. Retail sales of organic food by category, 2009. *Source:* Organic Trade Association.

As already mentioned, organic food sales accounted for 3.7 percent of all U.S. food sales in 2009. However, market penetration—the percentage of the overall market represented by organic food—was 9.5 percent for fruits and vegetables.

ORGANIC STRAWBERRY PRODUCTION IN CALIFORNIA

Information about the production of organic commodities in California is readily available because of the way organic producers, handlers, and processors in the state are required to register with the State Organic Program (SOP) of the California Department of Food and Agriculture. The information in their registration forms feeds a unique database that permits us to understand organic production in California beyond what is possible in other states. The 2007 U.S. Census of Agriculture included an organic follow-up survey for organic production and made those data available in 2008. Up until now, we have been talking about the organic food market in the United States in terms of food sold at the retail level, including processed foods. At this point we will begin a discussion of production in California based solely on sales at the farm gate.

Sales of organic commodities in California have grown at about 17 percent a year since 1991, when data were first collected (Klonsky et al. 2002; Klonsky and Richter 2005, 2011; Klonsky and Tourte 1998). Throughout that period, fruits and vegetables have accounted for more than 90 percent of organic sales, acreage, and farms. Strawberries are among more than 50 fruit crops that are grown organically in California. Sales have escalated, growing from $2 million in 1997 to over $55 million in 2009. Over that same period, acreage increased from 134 acres to 1,178 acres (table 1.1).

Table 1.1. Sales and acreage of organic strawberries in California, 1997–2009

Year	Sales ($ million)	Acres	Growers
1997	2.0	134	81
1998	8.3	235	83
1999	8.3	772	101
2000	9.7	509	122
2001	9.5	687	120
2002	12.8	1,250	111
2003	17.5	1,289	103
2004	31.2	1,987	111
2005	25.1	1,406	111
2006	30.1	1,868	119
2007	52.7	1,806	132
2008	67.6	2,701	153
2009	55.1	2,185	155

Strawberries are now produced organically in 32 counties in California—arguably, all of the counties in which the climatic conditions will support strawberry production. Almost two-thirds of growers reporting their organic strawberry sales to SOP were located on California's Central Coast or South Coast (table 1.2). In contrast, 97 percent of organic strawberry sales and acreage were from these two regions. This suggests that the largest organic strawberry growers are located on the coast, a pattern that is consistent with conventional strawberry production. The most important individual counties for organic strawberry production are Monterey, Santa Cruz, Santa Barbara, and Ventura. Together they accounted for roughly 90 percent of acreage and sales for organic strawberries in 2005 and were home to about half of all growers (table 1.3).

Table 1.2. Regional distribution of California organic strawberry sales, acres, and growers, 2005

Region	Sales ($)	Acres	Growers
Bay Area	57,085	10	14
Cascade	51,550	11	16
Central Coast	17,359,733	695	78
North Coast	151,163	16	30
Sacramento	92,180	5	6
San Joaquin Valley	27,500	3	5
South Coast	6,412,615	859	40
Southeast Interior	3,000	1	1

Table 1.3. Organic strawberry sales, acres, and number of growers for selected California counties, 2005

County	Sales ($ million)	Acres	Growers
Monterey	12.1	468	34
Santa Cruz	4.5	192	35
Santa Barbara	3.1	132	14
Ventura	2.5	641	8

REFERENCES/RESOURCES

Dimitri, C., and C. Greene. 2002. Recent growth patterns in the U.S. organic foods market. Washington, D.C.: USDA Economic Research Service, USDA Publication AIB-777. Accessed July 13, 2011, www.ers.usda.gov/Publications/aib777.

Klonsky, K., and K. Richter. 2005. A statistical picture of California's organic agriculture: 1998–2003. Davis: University of California Agricultural Issues Center. Accessed at http://aic.ucdavis.edu/research1/organic.html.

———. 2007. Statistical review of California's organic agriculture 2000–2005. Davis: University of California Agricultural Issues Center. Accessed at http://aic.ucdavis.edu/publications/Statistical_Review_00-05.pdf.

———. 2011. Statistical review of California's organic agriculture 2005–2009. Davis: University of California Agricultural Issues Center. Accessed at http://aic.ucdavis.edu/publications/Statistical_Review_05-09.pdf.

Klonsky, K., and L. Tourte. 1998. Statistical review of California's organic agriculture 1992–1995. Davis: University of California Agricultural Issues Center Statistical Brief No. 6 (May). Accessed at http://aic.ucdavis.edu/publications/oldanrpubs/organicstatistical.pdf.

Klonsky, K., L. Tourte, R. Kozloff, and B. Shouse. 2002. Statistical review of California's organic agriculture 1995–1998. Oakland: University of California Agriculture and Natural Resources. Publication 3425. Accessed at http://aic.ucdavis.edu/research/misc/Organic1995-98.pdf.

Nutrition Business Journal. 2010. NBJ Organic Report Featuring OTA Data: 2010. Accessed July 13, 2011, http://newhope360.com/agriculture/nbj-organic-report-featuring-ota-data.

USDA, NASS. 2008. 2008 Organic Production Survey, 2007 Census of Agriculture. Washington, D.C.: USDA National Agricultural Statistics Service.

PHOTO: ISTOCKPHOTO.COM

2 Costs of Production and Economic Returns for Organic Strawberries

LAURA TOURTE, KAREN KLONSKY, MARK BOLDA, AND RICHARD DE MOURA

ASSESSING COSTS OF ORGANIC PRODUCTION

The conversion of any farm to certified organic techniques can involve costs and risks. Organically grown strawberries can be profitable, though, in a well-managed fresh market operation. Currently, price premiums for organic strawberries are an important stimulus encouraging conversion and certification of farms, but it is difficult to predict the stability of premium prices over time, given market supply and consumer demand.

Net returns for organic strawberries can fluctuate in response to production and market conditions. It is therefore important that the grower make organic strawberry production decisions with full knowledge of local technical and marketing conditions. If you can devise reliable organic pest, soil, and field management systems for a given site, you can implement a transition management plan and then continually evaluate and update it, comparing its costs and revenues to those of conventional practices.

You can use a number of basic economic tools to analyze business decisions for an organic strawberry operation. These include whole farm budgets, enterprise budgets, partial budgets, and capital budgets. The type of budget you use in an analysis will depend on your business needs and the needs of your strawberry operation. Whole farm budgets are appropriate when you want to evaluate your entire business. Enterprise budgets are useful for determining the potential profitability of a specific commodity. Partial budgets enable you to examine the financial impact of specific changes within a cropping system—for example, changing pest management techniques or practices. Capital budgets are used to assess the impact of long-term business changes such as the purchase of land or equipment.

EXAMPLE: ESTIMATED COSTS AND RETURNS FOR CENTRAL COAST ORGANIC STRAWBERRIES

An enterprise budget for 2006, including a monthly cash flow analysis, for a 25-acre organic strawberry operation on the Central Coast (Santa Cruz and Monterey Counties) is presented in tables 2.1, 2.2, and 2.3. Costs are based on representative cultural practices, but these same practices are not used every year or by every grower. Costs, prices, and yields vary from farm to farm, within the season, and over time. Estimated cost tables are intended to serve as a baseline to help you understand your potential costs, make informed production decisions, determine your potential returns, prepare your own budget, and evaluate the business ramifications of production loans. Each actual farming situation is unique.

These are the underlying assumptions for the cost and return estimate:

- *Farm size.* The total farm size is 100 acres. Organic strawberries are planted on 25 acres, and 70 additional acres are planted to conventionally grown strawberries. The remaining acreage is reserved for roads and open space. The farmer rents the land at $2,000 per acre per year. This cost is within the range for strawberry rental land on the Central Coast for the 2006 study (Bolda et al. 2006).
- *Plant establishment and variety.* Strawberries are transplanted into 48-inch beds using two rows per bed at a 12-inch spacing, for a total plant density of 21,000 plants per acre. To account for plant loss during establishment, we assume that 5 percent of transplants are replanted. Several varieties are suitable for organic production in the region, including Seascape, Aromas, and Camarosa. However, no particular variety is specified for this study.

Table 2.1. 2006 operation costs ($/acre) for organic strawberries on the Central Coast of California*

Operation	Operation time (hr/acre)	Labor ($/acre)	Fuel and repairs ($/acre)	Materials ($/acre)	Custom/rent ($/acre)	Total ($/acre)
Cultural costs						
Land preparation: disk 2X	0.25	5	3	–	–	8
Fertilize: compost, blood meal	0.00	–	–	360	24	384
Irrigate: set up/remove sprinklers	3.00	87	24	–	–	111
Irrigate: sprinkler 2X	0.60	7	–	43	–	50
Land preparation: subsoil 2X	0.72	13	11	–	–	24
Land preparation: level land	0.16	3	2	–	–	5
Land preparation: disk 3X	0.37	7	5	–	–	12
Land preparation: rototill 2X	0.54	10	7	–	–	17
Land preparation: list/shape beds	0.23	4	3	–	–	7
Irrigate: install drip tape	1.06	30	9	240	–	279
Irrigate: open ditch for laterals	0.10	3	1	–	–	4
Plant: lay plastic mulch	0.52	10	4	363	–	377
Plant: open for transplants	0.52	10	4	–	–	14
Plant: transplant strawberries	100.00	1,035	–	1,654	–	2,689
Weed control: hand	200.00	2,070	–	–	–	2,070
Vertebrate control: trap	2.00	21	–	–	–	21
Weed control: cultivation	0.49	9	6	–	–	15
Fertilize: foliar applied	1.56	30	18	227	–	275
Irrigate: drip	12.00	124	–	650	–	774
Fertilize: drip applied	2.94	30	–	315	–	346
Arthropod pest control: mites	4.00	41	–	210	–	251
Arthropod pest control: worms	0.39	7	5	57	–	69
Disease control: powdery mildew	0.89	16	10	41	–	68
Pickup truck use	3.00	56	27	–	–	83
ATV use	1.02	19	2	–	–	21
Scouting services	0.00	–	–	–	150	150
Postharvest: cleanup	2.00	136	29	–	18	184
Total cultural costs	338.36	3,781	173	4,161	192	8,307
Harvest costs						
Harvest: regular	444.41	4,600	–	6,075	–	10,675
Harvest: haul to cooler	11.01	318	60	–	–	378
Harvest: piecework	0.00	–	–	–	3,677	3,677
Harvest: cooling	0.00	–	–	–	1,874	1,874
Assessments	0.00	–	–	300	–	300
Total harvest costs	455.42	4,918	60	6,375	5,551	16,904
Interest on operating capital @ 9.25%						1,131
Total operating costs/acre						26,342

Table 2.1. Continued

Operation	Operation time (hr/acre)	Labor ($/acre)	Fuel and repairs ($/acre)	Materials ($/acre)	Custom/rent ($/acre)	Total ($/acre)
Cash overhead costs						
Liability insurance						6
Office expenses						250
Sanitation expenses						43
Land rent (per producing acre)						2,105
Food safety program						8
Property taxes						17
Property insurance						12
Investment repairs						29
Total cash overhead costs						2,469
Total cash costs/acre						28,811

Investment	$/producing acre	Annual cost of capital recovery	Total ($/acre)
Noncash overhead costs	442	33	33
Buildings			
Fuel tanks	37	3	3
Miscellaneous tools	158	16	16
Harvest carts	12	3	3
Irrigation: pump & filters	158	14	14
Irrigation: sprinkler pipe	632	66	66
Irrigation: lateral lines	116	28	28
Equipment	1,469	156	156
Total noncash overhead costs	3,023	319	319
Total costs/acre			29,129
Total costs/tray			7.77

* Some rows and columns may not appear to add correctly, due to rounding.

Table 2.2. 2006 costs and returns ($/acre) for organic strawberries on the Central Coast of California*

	Quantity/acre	Cost ($)/unit	Cost ($)/acre
Gross returns organic strawberries	3,750 trays	11.00	41,250
Operating costs			
Fertilizer			
Compost	5 tons	27.00	135
Blood meal	500 lb	0.45	225
Biomin Ca 2-0-0-7	8 gal	14.00	112
Maxi Crop	20 lb	5.75	115
Liquid fertilizer 4-2-2	35 gal	2.55	89
Phytamin 801 (6-1-1)	35 gal	6.47	226
Spread fertilizer (compost and blood meal)	2.00 applications	12.00	24
Plants			
Strawberry transplants	22,050 plants	0.08	1,654
Irrigation			
Water, pumped	32 ac-in	8.33	267
Pajaro Valley Water Management Agency	32 ac-in	13.33	427
Drip tape	21,780 ft	0.01	240
Insect/mite control			
Predatory mites	40,000 mites	5.25/thousand	210
Dipel DF	4 lb	14.36	57
Scouting services	1	150.00	150
Disease control			
Micronized sulfur	45 lb	0.91	41
Misc. materials and fees			
Plastic mulch	2.75 rolls	132.00	363
Refuse disposal fee (mulch)	600 lb	0.03	18
Assessment fees			
Strawberry commission fees	3,750 trays	0.04	150
Organic certification/registration fees	1	150.00	150
Harvest			
Trays, clamshells	3,750 each	1.62	6,075
Piece rate per hour ($5.00 plus overhead)	234 hr	6.90	1,615
Piece rate per tray ($0.80 plus overhead)	1,875 trays	1.10	2,063
Cooling costs	3,750 trays	0.50	1,875
Labor (machine)	36 hr	15.46	553
Labor (nonmachine)	787 hr	10.35	8,146
Fuel, gas	19 gal	2.55	49
Fuel, diesel	45 gal	2.00	91
Lube			21
Machinery repair			72
Interest on operating capital @ 9.25%			1,131
Total operating costs/acre			26,342
Net returns above operating costs			14,908

* Some rows and columns may not appear to add correctly, due to rounding.

- *Yield.* Yield for organic strawberries varies depending on season and growing conditions. Yields range from 2,500 to 4,000 8-pound trays per acre. The assumed yield for this study is 3,750 8-pound trays per acre, which falls within the range of yields reported by growers in this area.
- *Labor.* Basic hourly wages for workers in the 2006 study are $11.20 for machine operators and $7.50 for fieldworkers. Adding 38 percent for workers' compensation, social security, and other benefits increases the labor rates to $15.46 per hour for machine labor and $10.35 per hour for field labor. During peak harvest months (June and July), pickers receive a piece rate calculated at $5.00 per hour plus $0.80 per tray. The labor hours for operations that involve machinery are 20 percent higher than actual operation time due to the extra time involved in equipment setup, moving, maintenance, work breaks, and repair.
- *Interest on operating capital.* This interest rate is based on cash costs for cultural and harvest operations. It is calculated monthly until harvest at the rate of 9.25 percent per year. You can view this interest as the cost of a production loan. For a business that is self-financed, you can see the interest as the opportunity cost of tying up capital in the crop.

Tables 2.1, 2.2, and 2.3 provide information for the hypothetical organic strawberry operation described in the set of assumptions. Each of the three tables presents the corresponding enterprise budget information in a different format, and as such each contains different levels of detail on production practices, inputs, and costs.

Table 2.1 lists the cash and labor costs according to the cultural operation used. The operations are listed in order of occurrence, and the cost of the material inputs is listed. Cash and noncash overhead costs are included to allow you to calculate total costs of production, excluding management and risk. Table 2.2 details the operating costs according to the inputs described by the operations in table 2.1. The total quantities and costs per unit of each input are presented. The cash and noncash overhead costs are not included here, to avoid redundancy with table 2.1. Table 2.3 lists the total operating costs by operation from table 2.1 on a monthly basis. The total cash costs per month are included.

For the hypothetical organic strawberry operation presented here, the total cash costs are estimated to be $28,811 per acre at a yield of 3,750 8-pound trays per acre. The total costs including capital recovery on buildings, harvest carts, well and pump, filters, equipment, and irrigation system are estimated at $29,129 per acre.

Total revenue per acre (gross returns) is calculated by multiplying the yield (number of trays per acre) by the price per tray. The price paid to growers typically ranges from $7 to $13 per 8-pound tray. Net returns represent revenue to growers minus a specific set of costs: for example, operating costs or cash costs. Net returns above total costs are considered to be profit, or returns to management and labor. One useful measure is to calculate the breakeven price and yield for the operation. For the breakeven price, this entails selecting an expected yield and then calculating what the price would have to be for the grower to receive revenue equal to the grower's costs at that expected yield. Similarly, to calculate the breakeven yield you would have to select an expected price and then calculate the yield at which total income would equal the cost of production.

For the farming operation studied here, assuming a yield of 3,750 trays per acre and prices at or below $7.77 per tray, it is not profitable to produce organic strawberries (table 2.1). At a price of $11 per 8-pound tray, the breakeven yield is calculated to be 1,907 trays per acre. Actual economic performance will depend on numerous factors including strawberry variety, site conditions, yearly production conditions, consumer demand, and market competition.

Cost of production studies for many crops, including other organic fresh market commodities, are available from the University of California Cooperative Extension. You can visit the UC Agricultural and Resource Economics Cost and Returns Studies website (http://coststudies.ucdavis.edu) or contact your local Cooperative Extension office. Studies are updated periodically, and analyses of additional crops are performed each year.

REFERENCES/RESOURCES

Bolda, M., L. Tourte, K. Klonsky, and R. De Moura. 2006. Sample costs to produce organic strawberries: Fresh market, Central Coast. Oakland: University of California Division of Agriculture and Natural Resources. Accessed at http://coststudies.ucdavis.edu.

Table 2.3. 2006 crop year monthly cash costs ($/acre) to produce organic strawberries on the Central Coast of California*

Beginning September 2005, ending October 2006	Sept 2005	Oct 2005	Nov 2005	Dec 2005	Jan 2006	Feb 2006	Mar 2006	Apr 2006	May 2006	June 2006	July 2006	Aug 2006	Sept 2006	Oct 2006	Total ($/acre)
Cultural costs															
Land prep: disk 2X	8	–	–	–	–	–	–	–	–	–	–	–	–	–	8
Fertilize: compost, blood meal	147	237	–	–	–	–	–	–	–	–	–	–	–	–	384
Irrigate: set up/remove sprinklers	55	–	55	–	–	–	–	–	–	–	–	–	–	–	111
Irrigate: sprinkler 2X	25	–	25	–	–	–	–	–	–	–	–	–	–	–	50
Land preparation: subsoil 2X	24	–	–	–	–	–	–	–	–	–	–	–	–	–	24
Land preparation: level land	5	–	–	–	–	–	–	–	–	–	–	–	–	–	5
Land preparation: disk 3X	12	–	–	–	–	–	–	–	–	–	–	–	–	–	12
Land preparation: rototill 2X	17	–	–	–	–	–	–	–	–	–	–	–	–	–	17
Land preparation: list/shape beds	–	7	–	–	–	–	–	–	–	–	–	–	–	–	7
Irrigate: install drip tape	–	279	–	–	–	–	–	–	–	–	–	–	–	–	279
Irrigate: open ditch for laterals	–	4	–	–	–	–	–	–	–	–	–	–	–	–	4
Plant: lay plastic mulch	–	377	–	–	–	–	–	–	–	–	–	–	–	–	377
Plant: open for transplants	–	14	–	–	–	–	–	–	–	–	–	–	–	–	14
Plant: transplant strawberries	–	2,689	–	–	–	–	–	–	–	–	–	–	–	–	2,689
Weed control: hand	–	–	–	207	207	207	207	207	207	207	207	207	207	–	2,070
Vertebrate control: trap	–	–	–	–	–	4	4	4	4	4	–	–	–	–	21
Weed control: cultivation	–	–	–	–	–	5	5	5	–	–	–	–	–	–	15
Fertilize: foliar applied	–	–	–	–	–	34	34	34	34	34	34	34	34	–	275
Irrigate: drip	–	–	–	–	–	–	110	110	111	111	111	110	110	–	774
Fertilize: drip applied	–	–	–	–	–	–	50	50	50	50	50	50	50	–	346
Arthropod pest control: mites	–	–	–	–	–	–	–	126	126	–	–	–	–	–	251
Arthropod pest control: worms	–	–	–	–	–	–	–	–	35	35	–	–	–	–	69
Disease control: powdery mildew	–	–	–	–	–	–	–	12	15	15	11	8	8	–	68

Pickup truck use	6	6	6	6	6	6	6	6	6	6	6	6	6	6	83
ATV use	2	2	2	2	2	2	2	2	2	2	2	2	1	1	21
Scouting services	—	—	—	—	—	—	150	—	—	—	—	—	—	—	150
Postharvest: cleanup	—	—	—	—	—	—	—	—	—	—	—	—	—	184	184
Total cultural costs	301	3,614	88	214	214	258	568	555	588	463	420	416	416	191	8,307
Harvest															
Harvest: regular	—	—	—	—	—	—	—	1,005	1,738	1,633	1,635	2,355	1,738	570	10,675
Harvest: haul to cooler	—	—	—	—	—	—	—	36	51	76	76	64	51	21	378
Harvest: piecework	—	—	—	—	—	—	—	—	—	1,838	1,839	—	—	—	3,677
Harvest: cooling	—	—	—	—	—	—	—	94	225	469	469	338	225	56	1,874
Assessments	—	—	—	—	—	—	—	—	—	—	—	—	—	300	300
Total harvest costs	—	—	—	—	—	—	—	1,136	2,015	4,016	4,020	2,757	2,015	947	16,904
Interest on operating capital	2	30	31	33	34	36	41	54	74	108	142	167	186	194	1,131
Total operating costs/acre	303	3,644	119	247	249	294	608	1,745	2,676	4,588	4,582	3,340	2,616	1,332	26,342
Cash overhead															
Liability insurance	—	—	—	—	6	—	—	—	—	—	—	—	—	—	6
Office expenses	21	21	21	21	21	21	21	21	21	21	21	21	—	—	250
Sanitation expenses	—	—	1	1	1	1	1	6	6	6	6	6	6	1	43
Land rent (per producing acre)	—	—	—	—	2,105	—	—	—	—	—	—	—	—	—	2,105
Food safety program	—	—	—	—	—	—	—	—	—	—	—	8	—	—	8
Property taxes	—	—	—	—	8	—	—	—	—	—	—	8	—	—	17
Property insurance	—	—	—	—	6	—	—	—	—	—	—	6	—	—	12
Investment repairs	2	2	2	2	2	2	2	2	2	2	2	2	—	—	29
Total cash overhead costs	23	23	25	25	2,150	25	25	29	29	29	29	51	6	1	2,469
Total cash costs/acre	327	3,667	143	272	2,399	319	633	1,774	2,705	4,616	4,610	3,391	2,622	1,334	28,811

* Some rows and columns may not appear to add correctly, due to rounding.

VARIETAL PERFORMANCE

PHOTO: KATHY KEATLEY GARVEY

VARIETAL PERFORMANCE

3 Cultivar Selection for Pest Management in Organic Strawberry Production

CAROLEE T. BULL AND FRANK N. MARTIN

In the future, advanced organic strawberry production will rely on technologies developed specifically for integration into the organic production system. The current organic strawberry production system in California is not optimized, primarily because many of the technologies currently in use are borrowed directly from conventional production with little or no adaptation to meet the needs of this distinct production system.

In particular, the cultivars used in California organic strawberry production are essentially the same as those developed to give optimal yields in conventional production systems. The widespread use of methyl bromide and chloropicrin soil fumigation has effectively controlled the most important soilborne pathogens for conventionally grown strawberries. This has led to the release of cultivars (some of which are widely planted in California) that are susceptible to many soilborne pathogens that are still present in nonfumigated and organically managed soils. In addition, these cultivars are optimized for production systems that depend on synthetic fertilizers, insecticides, and fungicides for plant nutrition and pest management. In organic production systems, cultivars must be capable of performing well under those fertilization and pest management strategies that are permitted by law, a significantly different set of options than are available for conventional production.

California organic strawberry farmers have a long list of high-quality, publicly available cultivars to choose from. Cultivars developed by the University of California strawberry improvement program are used throughout California and worldwide because of their high quality and high yields. Under conventional strawberry management, these cultivars are optimized for yield, production cycle, and fruit quality (Shaw and Larson 2008), and some of them have been evaluated for disease and insect resistance. Many of these cultivars perform well under a wide range of stress conditions (Shaw and Larson 1996, 2001, 2008). Organic strawberry growers rely on the data provided in cultivar release statements and in scientific publications to help them make informed choices of cultivars that will allow profitable production in an organic growing system.

Despite the high-quality cultivars available to organic strawberry growers in California, a growing body of evidence from research work in other crops suggests that cultivars bred specifically for organic production are needed if the growers are to optimize organic and low-input systems (Lammerts van Bueren et al. 1998, 2003; Murphy et al. 2005, 2007; Osman et al. 2008). For other crops, strong interactions have been detected between specific genotypes and organic or conventional production systems, indicating that further selection for suitability for organic fields would improve yields in organic systems (Boller, Tanner, and Shubiger 2008; Murphy et al. 2007). In these studies, certain plants performed better under organic production methods than might be expected, based on their performance in conventional production systems. Additionally, advanced tools such as quantitative trait loci (QTLs) are being applied by some to breeding in organic systems (Backes and Østergård 2008). Although most of these studies have been conducted on crops other than strawberry, there are indications of interactions between strawberry genotypes and the production systems in which they are grown (Shaw and Larson 2008) that plant breeders might exploit in order to advance cultivars that are intended specifically for use in organic production.

Breeding programs are expensive and labor intensive, and the move to breeding specifically for organic systems is proceeding at a slow pace. At a minimum, the selection of cultivars with characteristics important to organic production could result in systemwide improvements. Until organic-specific cultivars are developed, though, organic growers will have to continue to choose from among the existing publicly available cultivars and to rely on data from studies that are conducted mainly in conventional production fields.

Organic and conventional strawberry production systems differ significantly. Plant stresses differ due to differences in pressure from insects or insecticide applications, pathogens, and fungicide applications, and differences in available nutrition and organic matter. Soil fumigation with methyl bromide and chloropicrin, which is slowly being replaced by fumigation with other compounds (Ajwa et al. 2003; Fennimore et al. 2008), is perhaps the biggest difference between conventional and organic production practices. Soil fumigation reduces the number and diversity of microorganisms and weeds in the soil (Klose, Acosta-Martinez, and Ajwa 2006; Klose et al. 2007), killing both beneficial microorganisms and plant pathogens. In contrast, organic production practices are specifically designed to "feed the soil" and result in greater structural and functional diversity of soil microorganisms (Fließbach and Mäder 1997; Mäder et al. 2002). Although evaluations of cultivars in nonfumigated, conventionally managed production systems are not equivalent to those conducted in organic production systems, organic strawberry growers extrapolate the relative tolerance of cultivars to soilborne pests and weeds based on these studies in order to evaluate the potential performance of specific cultivars in organic systems. Cultivars that have been evaluated and selected in nonfumigated plots might be more useful in an organic setting than those that have been selected in fumigated soil, but they still are unlikely to be optimal for organic production because they still benefit from conventional synthetic pesticides and fertilizers.

In 2007, 63 percent of organically managed acres were planted to proprietary cultivars, compared to only 40 percent of conventionally managed acres. Several private breeding programs have developed cultivars from selections made in nonfumigated soils. California Giant, Inc., for example, has several proprietary cultivars (Cal Giant 2, 3, and 4) that were selected in and have performed well in nonfumigated soil. In conventional research plots that have been free of fumigation for 4 years, the best-performing cultivar was Cal Giant #4, which had an 18 percent reduction in yield in the nonfumigated treatment when compared to a fumigated treatment. Cultivars such as Camarosa and Diamante had 60 to 70 percent reductions in yield (figure 3.1).

Interestingly, the ranking of performance (based on yield) among cultivars for fumigated plots was different than the ranking for nonfumigated plots in these studies. Other studies have detected analogous interactions between genotype and crop production method (Shaw and Larson 2001, 2008), although ranking based on yield did not differ for fumigated vs. nonfumigated plots even after 7 years of repeated strawberry crops (Shaw and Larson 2001). Although the genotypes currently available are not optimized for organic production, a significant number of organic growers continue to rely on public breeding programs.

PEST RESISTANCE OF CULTIVARS

Diseases develop as the result of interactions between the host, a pathogen, and the environment. A change in state for any of these three can make a disease more or less likely to occur. Despite these interactions between the host, pathogen, and environment, no published research has yet evaluated disease resistance under organic management practices for California strawberry cultivars. Most of the information presented here was collected from conventional production fields, so organic growers must extrapolate from these data to make them apply to their own cultivar choices. There is a clear need for evaluations of the disease resistance and performance of currently available cultivars in organic fields.

Tolerance to a few soilborne plant pathogens has been reported for historical and current commercial cultivars, but no strawberry cultivar holds a monopoly on disease resistance (table 3.1). When considering the data on a cultivar's tolerance to specific pathogens, it is important to keep in mind that in the field the plants do not encounter a pathogen in isolation from other pathogens. Resistance to one pathogen does not ensure good performance in fields that have other pathogens.

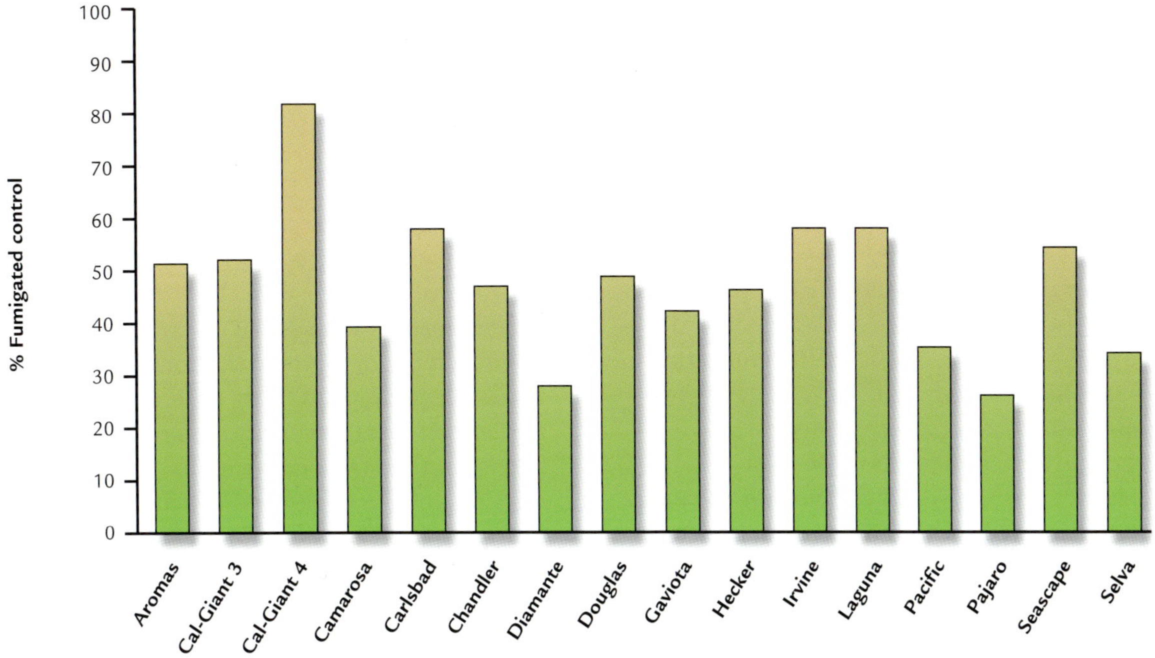

Figure 3.1. Cultivar performance in nonfumigated soil (expressed as a percentage of yield obtained in standard methyl bromide– and chloropicrin-fumigated soil in adjacent plots) in conventional field trials conducted at the USDA ARS research plots in Salinas in 2001.

Currently, only a few cultivars are tolerant to multiple pathogens. The incidence of disease resistance may increase, however, since public breeding efforts are now working on increasing cultivars' resistance to several soilborne diseases and insects.

Resistance to Verticillium Wilt

The soilborne disease Verticillium wilt, caused by the fungal pathogen *Verticillium dahliae*, is the most economically destructive disease of strawberries in nonfumigated conventional production. Currently, all commercially planted cultivars are susceptible to wilt, but the level of susceptibility differs among cultivars (table 3.1). For example, Duniway et al. (2001) found the cultivar Selva to be significantly less susceptible to Verticillium wilt than Camarosa when grown in conventional production fields. Other cultivars—namely, Pajaro, Capitola, Anaheim, Marmolada, and, again, Camarosa—were significantly more susceptible to Verticillium wilt than were Laguna, Seascape, Selva, and Chandler over several years of testing in conventional fields (Mircetich et al. 1994, 1995). Of these cultivars, only Camarosa continues to be grown organically or conventionally on any significant acreage. The more recently introduced Albion and Camino Real may be among the less-susceptible cultivars (Anonymous 2001). Camino Real has not been planted extensively, but Albion represented 12 and 29 percent, respectively, of the organic and conventional strawberry acreage in California in 2007 (Anonymous 2007). Ventana, which was grown on approximately 5 and 14 percent of organic and conventional acreage, respectively, is reported to be moderately susceptible to Verticillium wilt.

There are clear indications that it should be possible to develop increased levels of Verticillium wilt resistance in strawberries that would be effective against a broad range of isolates and that modern breeding material and cultivars already have some level of resistance (Gordon et al. 2006). Resistant lines were derived from clones of *Fragaria chiloensis*, *F. virginiana*, and *F. virginiana* subsp. *glauca*, which are at least moderately resistant to *Verticillium* spp. (Maas and Galletta 1989), although early studies did not identify resistance in commercial cultivars derived from the same species (Wilhelm 1955). Different levels of resistance to Verticillium wilt have been observed in field trials where the pathogen inoculum is applied by dipping the transplants into a conidial suspension prior to planting (Shaw, Gubler, and Hansen 1997), but before cultivar selection recommendations can be made to growers there must be additional field tests using natural microsclerotial inoculum of the pathogen. Additional research work intended to increase our understanding of the influence of cropping practices on the production of microsclerotia in organic systems is likely to yield additional useful management strategies.

Although Verticillium wilt is the most economically destructive disease in nonfumigated conventional strawberry production, its impact on organically managed strawberries has not yet been quantified. Crop rotation, organic amendments, and efforts to avoid the pathogen's introduction into fields where it has not previously been detected are important organic management strategies. Rotations with broccoli or Brussels sprouts have significantly reduced *V. dahliae* microsclerotia populations in soil and led to lower Verticillium wilt severity on strawberries in conventional (Subbarao et al. 2007) and organic (Njoroge et al. 2009) rotation plots. The benefits of broccoli rotations to organic strawberry production were confirmed on an additional organic farm (Muramoto et al. 2006). In some years and treatments, researchers detected an average of as many as 4 microsclerotia per gram of soil. These microsclerotia population levels fell significantly after incorporation of broccoli residues. However, levels of *Verticillium microsclerotia* and disease incidence were low in these plots, and no differences in this respect were detected between plots that had consecutive years of strawberry production and plots in which 4 years of vegetable production had preceded the planting of strawberries. See additional details on the effect of organic amendments on fertility and disease management in chapter 6, *Nitrogen Management for Organic Strawberries*, and chapter 9, *Cultural and Biological Management of Major Soilborne Diseases of Organic Strawberries*, and in P. D. Millner's 2006 article, "Control of Strawberry Black Root Rot with Compost Socks."

Resistance to Diseases Caused by *Phytophthora* spp.

Strawberry cultivars vary in their tolerance to root and crown diseases caused by the three different *Phytophthora* species (*P. cactorum, P. citricola,* and *P. fragariae*) encountered in California strawberry production (table 3.1; Browne et al. 1999, 2001). For example, Aromas was relatively tolerant to *P. cactorum* and *P. citricola,* whereas Diamante and Pajaro were more sensitive. The relative tolerance of Aromas made it rank as one of the best performing when cultivars were compared for their performance in organic production fields (Bull et al. 2005). *Phytophthora cactorum* is the dominant *Phytophthora* species in California conventional production systems, and data suggest that plants may transfer resistance to this pathogen to their offspring (Shaw et al. 2008). It appears that breeding strawberries for resistance to this pathogen may lead to improved performance in conventional production fields. Unfortunately, we do not know whether this species is similarly dominant in organic production fields (Gliessman et al. 1990; Rosado-May et al. 1994), though it would be very useful for organic strawberry growers to know which *Phytophthora* species were present in their soils and whether these pathogens could be expected to have a negative impact on organic strawberry production.

Soilborne Diseases Caused by Sublethal Pathogens

Different strawberry cultivars have shown varying degrees of response to some nonlethal root fungal pathogens—such as *Pythium, Cylindrocarpon,* and binucleate *Rhizoctonia* spp. (Martin 2000a, b)—that are associated with a root rot complex known as black root rot. These pathogens have been detected in organic production fields, but in the one study in which their effect was evaluated the populations were lower in organic fields than in comparable conventional fields (Rosado-May et al. 1994). In controlled greenhouse experiments, Aromas, Selva, Seascape, and Carlsbad were more tolerant to *Pythium ultimum* than were the other cultivars tested, including Torrey, Pajaro, Camarosa, Laguna, and Chandler (Martin 2000b). In contrast, binucleate *Rhizoctonia* spp. caused significant reductions in shoot growth and root weight for all cultivars examined.

Table 3.1. Relative resistance or susceptibility of California strawberry cultivars to soilborne diseases

	Cultivar response*		
Cultivar	**Tolerant**	**Intermediate**	**Susceptible**
Albion		Vert; Pcac†	
Anaheim			Vert‡
Aromas	Pcit; Pu§	Pcac; PRC	
Camarosa		Pcit; Pcac; PRC	Vert; Pu
Camino Real		Vert; Pcac	
Capitola	Pcit	PV, **PRC**¶	Vert
Carlsbad		PRC	
Chandler		Pu, Pu; PRC; **Pcit**	Vert; PV
Commander	Pcit		
Diamante	Pcit		Pcac; PRC
Douglas		PRC; Pcit	PV
Fern			Pcit
Gaviota		PRC	Pcac
Hecker		PRC	
Irvine		**PRC**	Vert; Pcit; PV
Laguna		PRC	Vert
Marmolada		Vert	
Muir			Pcit
Pacific		Pcac; PRC	
Pajaro		PRC; **Pcit**; Pff#	Vert; Pcac; PV
Parker	Pcac; Pcit	PRC	
Seascape		PRC	Vert; PV
Selva	Pu	PRC	Vert; PV
Sequoia		PRC	
Sheehy			Pcit
Tioga	Pcit		
Ventana		Pcac	Vert
Yolo			Pcit

* Tolerance was measured by comparing the yields or growth of the cultivars in the presence and absence of the pathogen. For all pathogens except *Pythium ultimum* (Pu), yields in infested soils were 90%, 50–89%, and less than 50% of the yields achieved in noninfested soil for tolerant, intermediate, and susceptible cultivars, respectively.

† Information for cultivars Albion, Camino Real, and Ventana are from the new cultivar release information available through the University of California.

‡ Pcac = *P. cactorum,* Pcit = *P. citricola,* Vert = *V. dahliae,* Pff = *P. fragariae* var. *fragariae,* Pu = *Pythium ultimum,* PRC = soil naturally infested with *Pythium, Cylindrocarpon,* and *Rhizoctonia* spp., and PV = soil infested with *Phytophthora* spp. and *V. dahliae.*

§ For Pu, results are based on greenhouse studies and ratings are based on fresh shoot weight (Martin 2000b).

¶ Pathogens in bold indicate variable results. Growers should look at the original data.

\# For Pff, all California cultivars evaluated were susceptible but the relative susceptibility was not given (Browne et al. 2001).

Sources: These data are based on greenhouse and field trials reported in Winterbottom et al. 1997; Mircetich et al. 1994, 1995; Browne et al. 1999, 2001; Martin 2000b, 2001; and Shaw, Larson, and Gubler 1994.

Although evaluations for resistance are conducted with one pathogen, in the field multiple pathogens can have a simultaneous influence on plant health and yield. For example, researchers conducted a field evaluation of cultivar performance in Salinas, California, in nonfumigated soil devoid of lethal pathogens (*V. dahliae* or *Phytophthora* spp.) but naturally infested with *Pythium* spp., binucleate *Rhizoctonia,* and *Cylindrocarpon* spp. The Aromas cultivar exhibited a 25 to 50 percent reduction in yield when grown in nonfumigated as compared to fumigated soil (figure 3.1), yet in the greenhouse this cultivar was very tolerant of *Pythium ultimum,* one of the predominant species at this site (Martin 2000b). This highlights the importance of conducting field evaluations to assess the performance of cultivars that have demonstrated resistance to one pathogen in greenhouse assays.

While these pathogens may have a significant impact on conventional strawberries grown in nonfumigated fields, it is not clear that they would have the same impact in organic production systems. Organically managed soils have greater microbial diversity than conventionally managed soils. Microbes in the soil or those associated with plant roots might be antagonistic to sublethal pathogens or might compete with pathogens for nutrients on the root surface or stimulate plant defenses. Sublethal pathogens are likely to be more sensitive to these microorganisms since they have been shown to be more amenable to control than lethal pathogens like *V. dahliae,* which causes Verticillium wilt.

A drop in yield for organic strawberries grown without rotation in fields with low disease pressure from major pathogens might indicate that sublethal pathogens are playing a significant role in reducing those yields. After 4 consecutive years of conventional strawberry production in the absence of fumigation and Verticillium wilt, strawberry yields in one test plot were observed to drop significantly (F. N. Martin, unpublished). This may have been the result of an increase in soilborne, sublethal fungal pathogens following a period of strawberry monoculture in nonfumigated plots. However, other research conducted on a working organic farm demonstrated that under low Verticillium pressure, fields using a combination of ecological practices that included the incorporation of broccoli residues could produce organic strawberries with a 1- to 3-year-break rotation without showing a significant difference in yield from a 7-year-break rotation (Muramoto et al. 2006). After 5 consecutive years of strawberry production, yields in these fields were significantly lower than for the 7-year-break rotation. The researchers indicated that the differences they detected might be the result of increases in nitrogen provided by nearly annual cover cropping in plots with a 7-year-break in strawberries, rather than a decrease in yield due to minor pathogens in the 5-year-consecutive strawberry plantings. These data further argue for the need for selection in authentic organic fields. Complex interactions between organic amendments and increased microbial diversity in soils may result in plant-pathogen-environment interactions that significantly alter the expression of disease and plant vigor.

RESISTANCE TO FOLIAR DISEASES AND PESTS

The active breeding of strawberry cultivars for resistance to foliar pathogens has received only limited attention in the United States because of the availability of chemical fungicides for control of the most important diseases in conventional production (Sorensen et al. 1997). However, in screening trials many breeding programs do not spray for foliar pathogens, and in this way they passively select for tolerance by not selecting individuals that show susceptibility. A few reports have referred to scientists' active evaluation of germplasm for resistance to foliar pathogens.

Of the cultivars evaluated under organic management for susceptibility to gray mold *(Botrytis cinerea),* powdery mildew *(Sphaerotheca alchemiillae),* and leaf spot *(Ramularia grevilleanan),* all but two were of European origin (Daugaard and Lindhard 2000). Honeoye, the only U.S. cultivar evaluated, was the least susceptible to *B. cinerea,* and the authors considered it to be a good cultivar for early-season organic production. Only one evaluation of foliar diseases has been reported in the literature in the United States. Nelson, Gubler, and Shaw (1996) determined powdery mildew resistance to be highly heritable and further determined that there is considerable potential for improvement of resistance. They hypothesized that increased resistance could help reduce growers' reliance on fungicides for powdery mildew control. Fortunately, the high variation for this trait in California genotypes means it is reasonable to expect that breeding programs will be able to improve resistance to powdery mildew. Their research may now be coming to fruition: the newer cultivars available for use in California differ from past cultivars in their resistance to foliar pathogens. Camino Real is relatively resistant to powdery mildew and anthracnose crown rot, while Albion and Ventana are relatively resistant only to anthracnose crown rot (www.innovationaccess.ucdavis.edu/strawberry/program.htm).

Insecticides have also precluded any need to breed for insect pest resistance in conventional strawberry production. Fortunately, tolerance to twospotted spider mite *(Tetranychus urticae)* is being evaluated in newer cultivars. Short-day cultivars are relatively tolerant to twospotted spider mite when planted in summer, but not when planted in the fall (Zalom, Phillips, and Toscano 2002). The newer cultivars Camino Real, Ventana, and Albion all are considered tolerant to twospotted spider mites when plants are carefully monitored and managed.

Although knowledge of predominant pests in organic production fields combined with the appropriate use of resistant cultivars may increase yields, resistance is not the only factor affecting plant performance. For example, the plants' ability to absorb nutrients from organic amendments may influence their performance (Bull et al. 2005). Organic breeding, including selection in organic production fields, would allow the cultivars to be optimized simultaneously for all factors present, and not just for those that are easy to test for and identify.

CULTIVAR PERFORMANCE IN ORGANIC PRODUCTION SYSTEMS

Organic growers used to base their choice of cultivar either on data extrapolated from research conducted under conventional production or on empirical data gained through their own experience. As an alternative to interpreting results from conventional research trials, growers can now draw on information from replicated trials on cultivar performance conducted in organic production fields (Bull et al. 2005), where researchers conducted five experiments over 3 years to evaluate cultivar performance in commercially managed certified organic production fields. Aromas, Pacific, and Seascape were the top-performing cultivars in the organic fields. Of these three, growers participating in these experiments preferred the flavor and market qualities of Seascape. Interestingly, in four of the five experiments, Diamante had significantly lower yields, even though at the time it was the dominant day-neutral cultivar planted in conventional production fields in the Central Coast region of California. With the exception of the Pacific cultivar, these data are consistent with the results from other trials conducted in conventional production fields without fumigation (figure 3.1).

CONCLUSIONS

To date, Aromas, Pacific, and Seascape have proven to be the best-performing publicly available cultivars for organic production fields in the Central Coast area of California (Bull et al. 2005). The supporting data are limited in that newer releases and proprietary cultivars were not included in the comparisons. The newer cultivars are likely to have better disease resistance because of new collaborations between plant pathologists and breeders to evaluate germplasm. For example, Aromas is at least tolerant, if not resistant, to *Phytophthora citricola, P. cactorum,* and *Pythium ultimum*. It also performed well in conventional fields infested with a mixture of *Pythium, Rhizoctonia,* and *Cylindrocarpon* spp. (Martin 2000a, 2000b). Resistance to these pathogens may be one factor that leads to better performance for Aromas under organic growing conditions. A strong ability to absorb nutrients (nitrogen in particular) from organic sources may be another factor contributing to high performance (Bull et al. 2005). Albion is now widely planted by organic growers. Essentially, all California cultivars are susceptible to *Verticillium dahliae,* but because the pathogen was not detected in soil at the test sites, it probably is safe to say that it had no effect on cultivars in the abovementioned trials.

California organic strawberry growers are fortunate in that they have a large number of cultivars commercially available to them. In addition, they can draw on a wealth of knowledge about the performance of cultivars in conventional production systems, both fumigated and nonfumigated, that can help them determine which cultivars are best suited to their planting situation. However, more cultivars specifically bred for organic production are becoming available to organic growers of other crops (Boller, Tanner, and Schubiger 2008; Lammerts van Bueren et al. 1998, 2003; Murphy et al. 2005, 2007; Osman et al. 2008).

Substantial resources are being allocated to meet organic-specific germplasm needs in Europe. A group of research institutes in Europe has established a consortium specifically focused on organic plant breeding (www.eco-ph.org). Breeders worldwide recognize that organic production systems differ significantly from conventional production systems and that the cultivars that work best for conventional systems are not always the same as those that work best for organic systems (Lammerts van Bueren et al. 1998, 2003; Murphy et al. 2005, 2007; Osman et al. 2008). This is probably true for strawberries, since there are indications that the relative performance of cultivars for some traits is dependent on the cropping system (F. N. Martin et al., unpublished; Shaw and Larson 2001, 2008). Although breeders at some U.S. companies are screening germplasm in nonfumigated soils, the authors have heard no reports of organic strawberry breeding as of this writing. Research and breeding at public institutions appears to be needed if we hope to advance organic strawberry production in California and the United States. Specifically, we need to develop cultivars that are optimized for organic production systems.

In addition to having cultivars that are specifically bred for organic strawberry production, growers need to know which pathogens are present in their organic production fields and at what populations, and what potential those pathogens have to contribute to reductions in their yields. Armed with this information, growers could make knowledge-based choices of which cultivar to plant. Further, an understanding of how increased microbial diversity affects plant health and how individual cultivar choices and organic farming practices impact the husbandry of beneficial insects could lead to significant improvements in organic production practices.

REFERENCES/RESOURCES

Ajwa, H. A., S. Klose, S. D. Nelson, A. Minuto, M. L. Gullino, F. Lamberti, and J. M. Lopez-Aranda. 2003. Alternatives to methyl bromide in strawberry production in the United States of America and the Mediterranean region. Phytopatholgia Mediterranae 42:220–244.

Anonymous. 2001. Strawberry licensing program. Accessed at http://innovationaccess.ucdavis.edu/strawberry/welcome.html.

Anonymous. 2007. CSC California strawberry 2007 acreage survey. Accessed at www.calstrawberry.com/commission/asurvey.asp.

Backes, G., and H. Østergård. 2008. Molecular markers to exploit genotype-environment interaction of relevance in organic growing systems. Euphytica 163:523–531.

Boller, B., P. Tanner, and F. X. Schubiger. 2008. Breeding forage grasses for organic conditions. Euphytica 163:459–467.

Browne, G. T., H. E. Becherer, M. R. Vazquez, S. A. McGlaughlin, R. J. Wakeman, C. Q. Winterbottom, J. M. Duniway, and S. A. Fennimore. 2001. Outlook for managing Phytophthora diseases on California strawberries without methyl bromide. Abstract 29 in Annual International Research Conference on Methyl Bromide Alternatives and Emissions Reductions. Accessed at http://mbao.org.

Browne, G. T., H. E. Becherer, M. R. Vazquez, and R. J. Wakeman. 1999. *Phytophthora* control on strawberry, almond, and walnut without methyl bromide. Abstract 82 in Annual International Research Conference on Methyl Bromide Alternatives and Emissions Reductions. Accessed at http://mbao.org.

Bull, C. T., J. Muramoto, S. T. Koike, J. Leap, C. Shennan, and P. Goldman. 2005. Evaluation of strawberry cultivars and mycorrhizal inoculants in California organic production fields. Crop Management doi:10.1094/CM-2005-0527-02-RS.

Daugaard, H., and H. Lindhard. 2000. Strawberry cultivars for organic production. Gartenbauwissenschaft 65:213–217.

Duniway, J. M., J. J. Hao, D. M. Dopkins, H. Ajwa, and G. T. Browne. 2001. Chemical, cultural, and biological alternatives to methyl bromide for strawberry. Abstract 41 in Annual International Research Conference on Methyl Bromide Alternatives and Emissions Reductions. Accessed at http://mbao.org.

Fennimore, S. A., J. M. Duniway, G. T. Browne, F. N. Martin, H. A. Ajwa, B. B. Westerdahl, R. E. Goodhue, M. Haar, and C. Q. Winterbottom. 2008. Methyl bromide alternatives for California strawberry nurseries. California Agriculture Journal 62:62–67.

Fließbach, A., and P. Mäder. 1997. Carbon source utilization by microbial communities in soils under organic and conventional farming practice. Pp. 109–120 in H. Insam and A. Rangger (eds.), Microbial communities: Functional versus structural approaches. Berline: Springer.

Gliessman, S. R., S. L. Swezey, J. Allison, J. Cochran, J. Farrell, R. Kluson, F. Rosado-May, and M. Werner. 1990. Strawberry production systems during conversion to organic management. California Agriculture Journal 44:4–7.

Gordon, T. R., S. C. Kirkpatrick, J. Hansen, and D. V. Shaw. 2006. Response of strawberry genotypes to inoculation with isolates of *Verticillium dahliae* differing in host origin. Plant Pathology 55:766–769.

Klose, S., V. Acosta-Martinez, and H. A, Ajwa. 2006. Microbial community composition and enzyme activities in a sandy loam soil after fumigation with methyl bromide or alternative biocides. Soil Biology and Biochemistry 38:1243–1254.

Klose, S., H. A. Ajwa, S. A. Fennimore, F. N. Martin, G. T. Browne, and K. V. Subbarao. 2007. Dose response of weed seeds and soilborne pathogens to 1,3-D and chloropicrin. Crop Protection 26:535–542.

Lammerts van Bueren, E. T., M. Hulscher, J. Jongerden, M. Haring, J. Hoogendoorn, J. D. van Mansvelt, and G. T. P. Ruivenkamp. 1998. Sustainable organic plant breeding. Dreibergen, The Netherlands: Louis Bolk Institute. Publication number G18.

Lammerts van Bueren, E. T., P. C. Struik, M. Tiemens-Hulscher, and E. Jacobsen. 2003. Concepts of intrinsic value and integrity of plants in organic plant breeding and propagation. Crop Science 43:1922–1929.

Maas, J. L., and G. J. Galletta. 1989. Germplasm evaluation for resistance to fungus-incited diseases. ACTA Hoticulturae 265:461–472.

Mäder, P., A. Fließbach, D. Dubois, L. Gunst, P. Fried, and U. Niggli. 2002. Soil fertility and biodiversity in organic farming. Science 296(5573):1694–1697.

Martin, F. N. 2000a. *Rhizoctonia* spp. recovered from strawberry roots on central coastal California. Phytopathology 90:345–353.

———. 2000b. Management of root diseases of strawberry. Abstract 32 in Annual International Research Conference on Methyl Bromide Alternatives and Emissions Reductions. Accessed at http://mbao.org.

———. 2001. Management of pathogens associated with black root rot of strawberry. Abstract 46 in Annual International Research Conference on Methyl Bromide Alternatives and Emissions Reductions. Accessed at www.mbao.org/2001proc/046%20Martin%20FN%20mbao%20-%2001%20abstract-b.pdf.

Millner, P. D. 2006. Control of strawberry black root rot with compost socks. Plant Health Progress doi:10.1094/PHP-2006-1016-02-RS.

Mircetich, J. S. M., D. Koball, R. J. Wakeman, C. Q. Winterbottom, F. Westerlund, L. Galper, N. T. Gargiulo, N. Welch, and W. D. Gubler. 1994. Comparative performance of different strawberry cultivars in field soil naturally infested by *Phytophthora* spp. and *Verticillium* spp. that was methyl bromide fumigated or nontreated. Abstract 45 in Annual International Research Conference on Methyl Bromide Alternatives and Emissions Reductions. Accessed at http://mbao.org.

Mircetich, J. S. M., C. Q. Winterbottom, R. J. Wakeman, F. Westerlund, L. Galper, N. T. Gargiulo, N. Welch, and W. D. Gubler. 1995. Comparative performance of different strawberry cultivars with and without methyl bromide fumigation in field soil naturally infested by *Phytophthora* spp. and *Verticillium* spp. and feasibility of using resistance of strawberry cultivars as an alternative to methyl bromide soil fumigation. Abstract 31 in Annual International Research Conference on Methyl Bromide Alternatives and Emissions Reductions. Accessed at http://mbao.org.

Muramoto, J., S. R. Gliessman, S. T. Koike, C. Shennan, D. Schmida, R. Stephens, and S. L. Swezey. 2006. Maintaining agroecosystem health in an organic strawberry/vegetable rotation system. Part 5: The final result. Annual Meeting Abstracts. Annual Meeting of the American Society of Agronomy, Indianapolis, Indiana, Nov. 12–16, 2006. Annual meetings abstract CD.

Murphy, K., D. Lammer, S. Lyon, B. Carter, and S. S. Jones. 2005. Breeding for organic and low-input farming systems: An evolutionary-participatory breeding method for inbred cereal grains. Renewable Agriculture and Food Systems 20:48–55.

———. 2007. Evidence of varietal adaptation to organic farming systems. Field Crops Research 102:172–177.

Nelson, M. D., W. D. Gubler, and D. V. Shaw. 1996. Relative resistance of 47 strawberry cultivars to powdery mildew in California greenhouse and field environments. Plant Disease 80:326–328.

Njoroge, S. M. C., Z. Kabir, F. N. Martin, S. T. Koike, and K. V. Subbarao. 2009. Comparison of crop rotation for Verticillium wilt management and effect on *Pythium* species in conventional and organic strawberry production. Plant Disease 93:519–527.

Osman, A. M., C. J. M. Almekinders, P. C. Struik, and E. T. Lammerts van Bueren. 2008. Can conventional breeding programmes provide onion varieties that are suitable for organic farming in the Netherlands? Euphytica 163:511–522.

Rosado-May, F. J., M. R. Werner, S. R. Gliessman, and R. Webb. 1994. Incidence of strawberry root fungi in conventional and organic production systems. Applied Soil Ecology 1:261–267.

Shaw, D. V., W. D. Gubler, and J. Hansen. 1997. Field resistance of California strawberries to *Verticillium dahliae* at three conidial inoculum concentrations. Journal of the American Society for Horticulture Science 32:711–713.

Shaw, D. V., J. Hansen, G. T. Browne, and S. M. Shaw. 2008. Components of genetic variation for resistance of strawberry to *Phytophthora cactorum* estimated using segregating seedling populations and their parent genotypes. Plant Pathology 57:210–215.

Shaw, D. V., and K. D. Larson. 1996. Relative performance of strawberry cultivars from California and other North American sources in fumigated and non-fumigated soils. Journal of American Society for Horticulture Science 121:764–767.

———. 2001. Relative performance of strawberry genotypes over four cycles of cultivation on fumigated and nonfumigated soils. Journal of American Society for Horticulture Science 126:78–82.

———. 2008. Performance of early generation and modern strawberry cultivars from the University of California breeding programme in growing systems simulating traditional and modern horticulture. The Journal of Horticulture Science and Biotechnology 83:648–652.

Shaw, D. V., K. D. Larson, and W. D. Gubler. 1994. An evaluation of the potential for developing strawberry cultivars adapted to nonfumigated soils. Abstract 44 in Annual International Research Conference on Methyl Bromide Alternatives and Emissions Reductions. Accessed at http://mbao.org.

Sorensen, K. A., W. D. Gubler, N. C. Welch, and C. Osteen. 1997. The importance of pesticides and other pest management practices in U.S. strawberry production. National Agricultural Pesticide Impact Assessment Program. Document Number 1:CA–97.

Subbarao, K. V., Z. Kabir, F. N. Martin, and S. T. Koike. 2007. Management of soilborne diseases in strawberry using vegetable rotations. Plant Disease 91:964–972.

Wilhelm, S. 1955. *Verticillium* wilt of the strawberry with special reference to resistance. Phytopathology 45:387–391.

Winterbottom, C. Q., F. Westerlund, J. Mircetich, and L. Galper. 1997. Evaluation of relative resistance of different strawberry cultivars to *Phytophthora* and *Verticillium dahliae* as a potential alternative to methyl bromide. Abstract 33 in Annual International Research Conference on Methyl Bromide Alternatives and Emissions Reductions. Accessed at http://mbao.org.

Zalom, F. G., P. A. Phillips, and N. C. Toscano. 2002. UC IPM pest management guidelines: Strawberry, insects, and mites. Oakland: University of California Division of Agriculture and Natural Resources. Publication 3468.

CROP MANAGEMENT

PHOTO: KATHY KEATLEY GARVEY

4

Cover Crops for Nutrient Management in Organic Strawberry Production on California's Central Coast

RICHARD SMITH

Cover cropping is an age-old farming practice that can provide important benefits to organic strawberry production. Cover crops can benefit soils and crop production by increasing soil nitrogen (N) availability, conserving nutrients (i.e., reducing nitrate leaching), reducing soilborne pest pressure, improving soil physical properties, and reducing soil erosion (Miller et al. 1989). In this chapter we will discuss the role of cover crops in crop production and address their impact on plant nutrition for organic strawberries.

An organic production system must employ soil-building practices, which can be summarized as feeding the soil to feed the plant by maintaining adequate levels of available nutrients in the soil nutrient pool (Gaskell et al. 2006). Growers can apply this precept by means of cover cropping and other practices that increase soil organic matter, soil biological activity, and nutrient availability. Cover cropping is a key component in this soil-building process. Cover crops can be a practical, economical means for increasing the soil's organic matter content and maintaining its productivity (Miller et al. 1989). Organic matter improves the tilth of the soil and serves as a reservoir for nutrients that will become available to succeeding crops as the organic matter decomposes. Organic matter also provides a food source for beneficial soil organisms that improve soil health (Mitchell et al. 2000).

Along with the benefits of cover crops, though, there are also potential disadvantages. One is the loss of potential income when you grow a cover crop instead of a cash crop. This gives growers a difficult choice in the strawberry growing regions of the Central Coast, where land rents often exceed $2,000 per acre (Bolda et al. 2006). The production costs of growing and incorporating a cover crop can also be significant (Tourte et al. 2003). In addition, under some conditions a cover crop can significantly increase weed problems if the weeds set seed during the cover-cropping period (Brennan and Smith 2005). Because weed control in organic strawberry production is so expensive, it is important to minimize weed seed production in all phases of the crop rotation.

COVER CROP GROWING SEASONS

Cover crops can be grown year-round on the Central Coast, but for the most part they are planted in three seasonal time slots: winter (October to March; 120 to 150 days to mature), summer (April to August; 45 to 75 days to mature), and fall (August to November; 45 to 90 days to mature). Cover crops are usually grown the winter or summer before you plant strawberries. The choice of growing season will influence both your choice of cover crop and its relative productivity.

TYPES OF COVER CROPS

Most of the cover crops suggested here are cool-season annuals, unless otherwise noted. They are usually grown in the winter, but most can be grown year-round in the mild climate of the strawberry production areas of coastal California.

Grasses

Grass cover crops include cereal rye (*Secale cereale*), barley (*Hordeum vulgare*), oats (*Avena sativa*), triticale (× *Triticosecale rimpau*), and wheat (*Triticum aestivum*), as well as Sudangrass (*Sorghum sudanense*) and sorghum-Sudangrass (*S. bicolor* [syn. *S. vulgare*] × *S. sudanense*) (warm-season annuals). There are many varieties of each of these species, each with its particular strengths and weaknesses. Merced rye is a popular cover crop on the Central Coast because it is productive, has an extensive root system, is late maturing, is tolerant of poor soil conditions, and has a low cost for seed (figure 4.1). High productivity is desirable because it determines the amount of biomass the cover crop will add to the soil. Late maturity is important because if the cover crop sets seed too early, its seedlings can become weeds in subsequent cash crops. This is particularly important when you grow early cover crops over the winter and late spring rains prevent your timely incorporation of the cover crop into the soil prior to seed set. Merced rye and some varieties of oat (e.g., Cayuse and Pert) are examples of late-maturing cover crops, whereas most barley varieties are early maturing. If you use early maturing cover crops, it is safer to plant them in the summer planting slots so you will be able to incorporate them into the soil before rainy weather can hamper timely field operations.

Figure 4.1. Young seedlings of Merced rye drill planted as a solid stand. *Photo* by R. Smith.

Cool-season grasses are the most cold-tolerant cover crops and can be planted in the coldest months (December and January). However, varieties such as Merced rye and Cayuse oats still benefit from earlier planting dates that allow them to mature early enough in the following spring to provide reasonable biomass production (table 4.1). Most cereal varieties are also reasonably competitive with weeds when sown at or above commonly recommended seeding rates. Warm-season species such as Sudangrass and sorgum-Sudangrass can be planted spring through summer. During this time of year, there is no rainfall, and they require irrigation to complete their growth cycle.

Legumes

Fast-growing legume cover crops include bell bean (*Vicia faba*), vetches (*Vicia* spp.), and peas (*Pisum sativum*). Legumes are important in organic systems because they form a symbiotic association in their roots with a soil microbe (*Rhizobium* sp.) that, when growing in association with the legume, can convert nitrogen in the air (dinitrogen: N_2), which is not available to plants, into a form of nitrogen that can be utilized by plants (ammonium: NH_4). This is called nitrogen fixation. Over a 6-year period, legume cover crops on the Central Coast contained an average of 146 pounds of nitrogen per acre in the tops (table 4.2). The legumes generally produce moderate amounts of biomass (e.g., 2 ton/acre), but in wet years or on poorly drained soils, some species are susceptible to soilborne and foliar diseases that can reduce their productivity. Legumes are sensitive to cold temperatures and grow slowly during the winter if planted too late in the fall (e.g., after October). They are also less competitive with weeds than are grass cover crops and can allow significant amounts of weed seed production under the cover crop canopy when they are grown in the winter. For instance, in a 2-year study conducted in Salinas, weed seed production was twice as high under a legume-cereal mix than under an oat cover crop, and ten times greater than under a mustard cover crop (Brennan and Smith 2005). It is possible to reduce the tendency of legume mixes to allow unchecked establishment and seed set by weeds if you use higher seeding rates (e.g., 200 lb/acre).

Mustards

Mustard cover crops include white mustard *(Sinapis alba)*, Indian mustard *(Brassica juncea)*, and canola *(B. napus)*. Upon incorporation into the soil, mustard cover crops release short-lived toxic chemicals (glucosinolates), which can biofumigate the soil. Some mustard cover crops have been found to reduce populations of soilborne nematodes (Potter, Davies, and Rathjen 1998) and weeds (Bensen et al. 2009). Broccoli *(B. oleracea)* is also in the mustard group, and rotations with broccoli have been shown to reduce the incidence of Verticillium wilt in subsequent crops (Shetty et al. 2000).

Mustards are most productive if planted in the fall. White mustard sets seed relatively early, and you must take care with overwintered plantings to avoid letting the plants set seed before you have a chance to incorporate them into the soil in spring. Mustard cover crops generally maintain a high concentration of nitrogen in the leaf tissue and release nitrate-nitrogen into the soil, something like what legumes do (figure 4.2).

Table 4.1. Biomass, percentage nitrogen (N) in tops, and total N in tops for three cover crops planted on three dates and evaluated on two dates

Cover crop – Planting date	Biomass (ton/acre)		Percentage N in tops		Total N in tops (lb/acre)	
	March 1	March 26	March 1	March 26	March 1	March 26
Barley – Oct. 23	3.19	4.47	1.87	1.31	120.0	121.0
Barley – Nov. 15	1.32	2.10	2.38	1.65	66.9	72.5
Barley – Dec. 15	0.16	0.51	2.61	2.43	14.7	24.5
Merced rye – Oct. 23	4.15	5.07	1.69	1.28	136.1	129.3
Merced rye – Nov. 15	1.75	4.64	2.29	1.34	77.7	145.3
Merced rye – Dec. 15	0.15	0.51	1.97	1.88	8.0	18.0
Legume/cereal mix* – Oct. 23	3.40	3.62	3.89	3.06	266.2	219.6
Legume/cereal mix* – Nov. 15	1.49	2.74	3.64	3.63	106.3	202.4
Legume/cereal mix* – Dec. 15	0.10	1.13	4.17	4.14	16.9	92.2

* Contained Cayuse oats, bell beans, Magnus pea, Lana woollypod vetch, and common vetch.

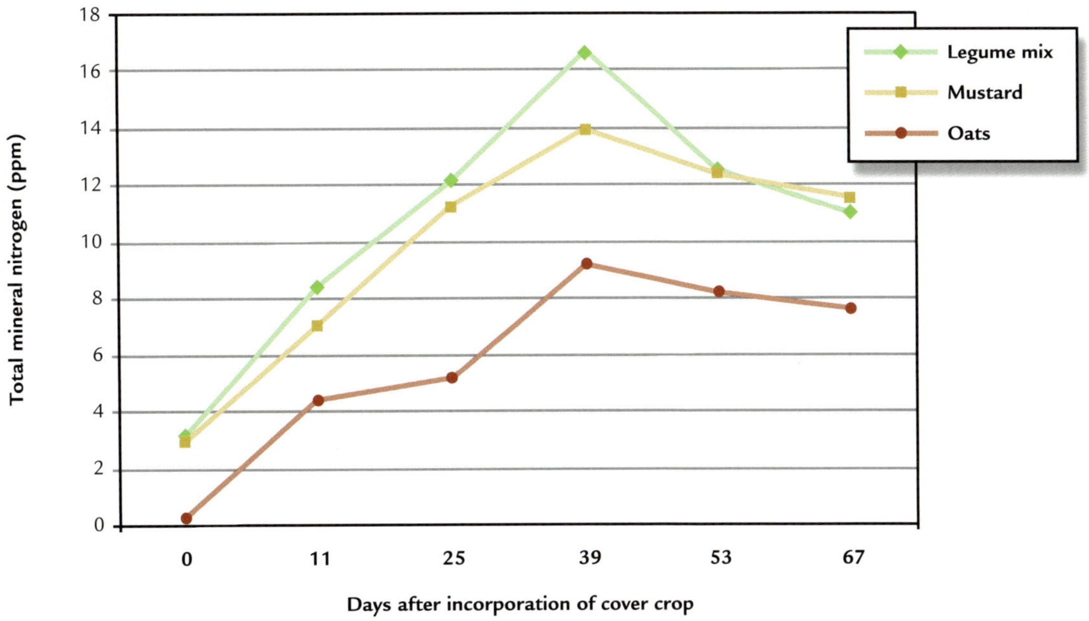

Figure 4.2. Total mineral nitrogen in the soil in three cover crop types on six dates following incorporation of cover crop into the soil (Smith and Brennan, 2004, unpublished data).

Other Cover Crops

Other cover crop options on the Central Coast include buckwheat and phacelia. In coastal regions, buckwheat *(Fagopyrum esculentum)* is a fast-growing, warm-season annual that can complete its life cycle and set viable seed in as few as 50 to 60 days. Phacelia *(Phacelia tanacetifolia)* is a vigorous, cool-season annual that is relatively productive and easy to incorporate into the soil. Warm-season legumes such as cowpeas *(Vigna unguiculata* ssp. *unguiculata)* do not produce well in the cool climate of the Central Coast.

Figure 4.3. Stand of a cover crop mix consisting of Merced rye and vetch. *Photo* by R. Smith.

COVER CROP MIXTURES

Mixtures of cover crops can provide multiple benefits to the soil-plant system and may provide insurance against conditions that are unfavorable to a single species (figure 4.3). In legume and cereal mixes, the cereals can absorb residual nitrogen from the soil while the legumes provide a net input of nitrogen to the soil by means of nitrogen fixation. By including multiple legume species and varieties in a cover crop planting, you may help ensure optimal nitrogen fixation. Such mixes were consistently productive over a 6-year period, and on average contained more nitrogen than cover crops made up of single legume species or of grass crops alone (table 4.2).

SOIL NITROGEN

Cover crops play a key role in providing nitrogen for organic crop production. The nitrogen contained in cover crops becomes available to subsequent crops as the cover crop decomposes in the soil. In this process, more complex forms of nitrogen (e.g., protein) in the cover crop tissue break down into plant-available forms (e.g., NH_4 and NO_3). Decomposition is carried out by soil microbes and proceeds more quickly in warmer soils with adequate soil moisture, with cover crops that have a higher nitrogen content, and with cover crops that have been cut into smaller pieces (e.g., mowed residue as opposed to unmowed residue).

Leguminous cover crops can be particularly rich in nitrogen. When they fix atmospheric nitrogen, they provide a net increase in soil nitrogen. This is important because nitrogen is a significant expense for organic producers. For example, the current cost for nitrogen from conventional fertilizers is less than $0.50 per pound. However, the cost of nitrogen from dry organic fertilizers is over $2.50 per pound, and costs are much higher for liquid organic fertilizers.

A legume cover crop can provide an economical source of nitrogen. For instance, legume cover crops typically cost $100 to $150 per acre to grow. If they produce 100 to 150 pounds of nitrogen per acre, the cost of the nitrogen contained in cover crops (approximately $1.00/lb of nitrogen) is cheaper than the cost of purchased inputs. Particularly productive legume varieties such as Lana woollypod vetch and Magnus pea can produce more than 200 pounds of nitrogen per acre in pure stands in some years (table 4.2). Bear in mind, though, that these estimates of the cost of cover crop nitrogen do not include land rents or the opportunity costs of growing a cover crop. Still, the net input of nitrogen from cover crops can satisfy a significant amount of the early-season nitrogen needs of strawberries.

Cereal cover crops do not fix nitrogen from the air, but their extensive root systems are capable of scavenging large quantities of nitrogen from the soil, maintaining it in the soil-plant system, and cycling it to the subsequent crop after soil incorporation (Wyland et al. 1996). In soils with a high level of residual nitrogen, cereal cover crops can absorb 150 to 200 pounds of nitrogen per acre into their tops (table 4.2).

The nitrogen contained in cover crops eventually becomes available to subsequent crops, but the rate at which it becomes available depends on the cover crop species, its maturity, and the nitrogen content of its tissue. In general, nitrogen is released most quickly from legumes, then from mustards, and then from cereals (figure 4.2). Cover crops with higher nitrogen

Table 4.2. Summary of cover crop biomass and nitrogen (N) content from 1995 to 2000, San Benito County

Variety	1995	1996	1997	1998	1999	2000	6-year mean
				biomass (ton/acre)			
Legumes							
Bell bean	2.3	5.0	2.2	3.0	–	1.8	2.8
Lana woollypod vetch	1.3	2.4	2.1	2.3	1.4	1.9	1.9
Purple vetch	–	2.7	1.4	2.4	0.5	2.3	1.9
Languidoc vetch	–	–	0.4	3.8	0.3	2.0	1.6
Magnus pea	2.7	3.8	0.2	2.2	0.2	2.2	1.9
All legumes							2.0
Cereals							
Merced rye	–	–	3.8	2.2	4.3	4.2	3.6
Montezuma oats	–	4.9	3.5	2.4	–	–	3.6
Cayuse oats	5.1	6.5	3.8	4.3	3.4	2.7	4.3
All cereals							3.8
Cereal/legume mix	3.9	–	2.6	3.0	3.3	2.5	3.1
				total N in tops (lb/acre)			
Legumes							
Bell bean	118.0	247.0	133.2	142.9	–	125.4	153.3
Lana woollypod vetch	120.0	192.0	214.4	203.0	130.7	207.6	177.9
Purple vetch	–	175.0	98.0	163.8	43.7	221.5	140.4
Languidoc vetch	–	–	27.9	249.8	26.4	171.5	118.9
Magnus pea	210.0	246.0	14.3	143.6	17.1	197.6	138.0
All legumes							145.7
Cereals							
Merced rye	–	–	85.2	101.6	198.9	167.1	138.2
Montezuma oats	–	119.0	108.9	93.9	–	–	107.3
Cayuse oats	195.0	127.0	101.2	139.7	198.7	133.7	149.2
All cereals							131.5
Cereal/legume mix	215.0	–	105.6	148.9	216.5	199.3	177.1

concentrations (>2.5%) at the time of incorporation will be quicker to release their nitrogen into the soil. Cover crops with lower nitrogen concentrations will release it more slowly.

Cover crop residue that is low in nitrogen (<1.5%) may immobilize its nitrogen for a period of time following incorporation. This is because soil microbes have to use available soil nitrogen to decompose the cover crop residue itself (Williams et al. 1968). During cover crop decomposition, a portion of the cover crop's nitrogen becomes incorporated into soil humus. That nitrogen may become available for crop growth in later growing seasons (Clark et al. 1998). Exactly how much nitrogen will be released from a cover crop is difficult to say in an organic system because it is dependent upon a complex interaction between soil microbes and the processes of organic matter formation and degradation. These processes are poorly understood, and currently there is no reliable test for predicting the rate of nitrogen release from soil organic matter (Horwath and Pettygrove 2006).

In addition, any synchronization of the release of this nitrogen with the needs of the crop is difficult to achieve. Studies on coastal vegetables indicate that the available nitrogen from a cereal-legume mix of cover crops is released over a 6- to 10-week period following incorporation into moist soil (Gaskell and Smith 2007). Following the peak of release, soil nitrogen levels return to pre-incorporation levels.

Gaskell (2004) observed that a cereal-legume mix cover crop provided about 100 pounds of nitrogen for subsequent pepper and cabbage crops. No studies have been conducted on the equivalent fertilizer value of cover crops for organic strawberries. While studies on other crops do provide a measure of the amount of nitrogen potentially available from recently incorporated cover crop residue, they can only provide a rough estimate for strawberry growers since the growth and nitrogen demand patterns for strawberries are very different from those of vegetable crops. In addition, the eventual fate of cover crop nitrogen can be affected by environmental factors. For instance, significant precipitation between the time of incorporation and the time of crop planting can leach nitrogen out of the soil, reducing the amount of cover crop nitrogen that will even reach the cash crop (Muramoto et al. 2011).

In most instances, cover crop residues may not provide sufficient nitrogen to meet the strawberry crop's needs and achieve high yields. Strawberries are a long-season crop that benefits both from initial and sustained release of cover crop nitrogen (see chapter 6, *Nitrogen Management for Organic Strawberries*). For high yields, you will need to make supplemental applications of nitrogen beyond the amount provided by cover crops, especially later in the growing season when the release of nitrogen from cover crops has substantially declined (Gaskell 2009). Over a period of years, however, the soil organic matter will build up from the use of cover crops and other organic amendments (i.e., compost), and the amount of supplemental fertilizer nitrogen needed to achieve high strawberry yields may eventually be reduced.

Figure 4.4. Trios 102 triticale growing in furrows of young strawberry planting. *Photo* by R. Smith.

Cover crops are also important for cycling other key nutrients including phosphorus and potassium. The forms of phosphorus and potassium contained in the cover crop residue are more readily available to plants than those contained in the mineral fraction of many soils.

IN-FURROW COVER CROPS

Cover crops can be grown in the furrows of strawberry plantings to provide protection against erosion on sloping ground during the winter (figure 4.4). This technique is distinct from the cover crop culture we have discussed up until now, where the cover crop is grown and incorporated into the soil prior to planting strawberries. In-furrow cover crops are particularly useful on beds with plastic mulch because they can have such high rates of runoff. In studies conducted near Watsonville, these cover crops reduced total sediment loss by more than 80 percent (figure 4.5) (Smith, Cahn, and Bolda 2007). Many cover crop species are suited to this technique; Trios 102 triticale, however, is winter dormant and requires less maintenance than other varieties that grow vigorously during the winter months. The height of in-furrow cover crops must be managed to keep them from shading out strawberry plantings. Organic in-furrow cover crop management includes weed whacking, cultivating, and trampling. If properly managed, in-furrow cover crops can provide erosion protection and reduce nutrient loss and still have no negative effect on the strawberry yield.

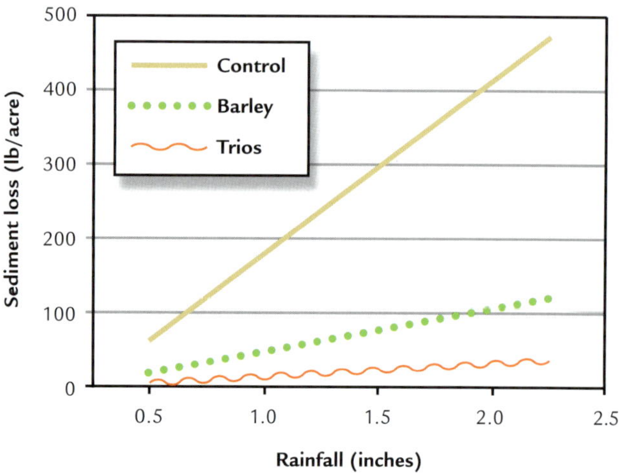

Figure 4.5. Cumulative sediment loss in runoff from in-furrow cover crops in strawberries (collected over 5 weeks during winter storms, 2007).

CONCLUSIONS

Cover cropping is an important tool for increasing and maintaining organic matter content and available nutrients in the soil. Decomposing cover crops increase the available nitrogen in the soil, both short- and long-term. Mineralized nitrogen from cover crops can satisfy a large part of the nitrogen needs of organic strawberries, but given strawberries' long production season, supplemental nitrogen fertilizer applications are nececessary to meet the crop's nitrogen need and achieve high yields.

The complex nature of the release of nitrogen from cover crops and the interaction of soil physical and biological characteristics make it hard to predict just how much nitrogen will be available from cover crops and when. Studies are needed on nitrogen from cover crops in organic strawberry production to help explain cover crops' role providing nitrogen for crop growth.

REFERENCES/RESOURCES

Bensen, T. A., R. F. Smith, K. V. Subbarao, S. T. Koike, S. A. Fennimore, and S. Shem-Tov. 2009. Mustard and other cover crop effects vary on lettuce drop caused by *Sclerotinia minor* and weeds. Plant Disease 93(10):1019–1027.

Bolda, M., L. Tourte, K. Klonsky, and R. De Moura. 2006. Sample costs to produce organic strawberries. Oakland: University of California Division of Agriculture and Natural Resources. Accessed at http://coststudies.ucdavis.edu.

Brennan, E. B., and R. F. Smith. 2005. Winter cover crop growth and weed suppression on the Central Coast of California. Weed Technology 19:1017–1024.

Clark, M. S., W. Horwath, C. Shennan, and K. Scow. 1998. Changes in soil chemical properties resulting from organic and low-input farming practices. Agronomy Journal 90:662–671.

Gaskell, M. 2004. Nitrogen availability, supply, and sources in organic row crops. In Proceedings of the 4th Annual California Conference on Biological Control, Berkeley, July 15, 2004.

———. 2009. Field evaluations of liquid fertilizers on organic strawberries. Presentation at the 2009 Irrigation and Nutrient Management Seminar. http://cemonterey.ucdavis.edu/files/63094.pdf.

Gaskell, M., and R. F. Smith. 2007. Nitrogen sources for organic vegetable production. HortTechnology 17(4):431–441.

Gaskell, M., R. F. Smith, J. Mitchell, S. T. Koike, C. Fouche, T. Hartz, W. Horwath, and L. Jackson. 2006. Soil fertility management for organic crops. Oakland: University of California Division of Agriculture and Natural Resources. Publication 7249.

Horwath, W. R., and G. S. Pettygrove. 2006. Practical soil test methods for predicting N mineralization. In Proceeding of the 14th Annual Fertilizer Research and Education Program Conference, Seaside, November 29, 2006.

Miller, P. R., W. L. Graves, W. A. Williams, and B. A. Madson. 1989. Covercrops for California agriculture. Oakland: University of California Division of Agriculture and Natural Resources. Publication 21471.

Mitchell, J., M. Gaskell, R. Smith, C. Fouche, and S. Koike. 2000. Soil management and soil quality for organic crops. Oakland: University of California Division of Agriculture and Natural Resources. Publication 7248.

Muramoto, J., R. Smith, C. Shennan, K. Klonsky, J. Leap, M. Silva Ruiz, and S. Gliessman. 2011. Nitrogen contribution of legume/cereal mixed cover crops and organic fertilizers to an organic broccoli crop. HortScience 46(8):1154–1162.

Potter, M. J., K. Davies, and A. J. Rathjen. 1998. Suppressive impact of glucosinolates in *Brassica* vegetative tissues on root lesion nematodes *Pratylenchus neglectus*. Journal of Chemical Ecology 24(1):67–79.

Shetty, K. G., K. V. Subbarao, O. C. Huisman, and J. C. Hubbard. 2000. Mechanism of broccoli-mediated Verticillium wilt reduction in cauliflower. Phytopathology 90(3):305–310.

Smith, R. F., M. Cahn, and M. Bolda. 2007. Evaluation of cover crops to improve water quality in vegetable and strawberry production. Final report to the Monterey Community Foundation, grant 20040584.

Tourte, L., M. Buchanan, K. Klonsky, and D. Mountjoy. 2003. Central coast conservation practices: Estimated costs and potential benefits for an annually planted cover crop. University of California Cooperative Extension. Santa Cruz County, California. Accessed at http://coststudies.ucdavis.edu/conservation_practices.

Williams, W. A., D. S. Mikkelsen, K. E. Mueller, and J. E. Ruckman. 1968. Nitrogen immobilization by rice straw incorporated in lowland rice production. Plant and Soil 28:49–60.

Wyland, L. J., L. Jackson, W. Chaney, K. Klonsky, S. Koike, and B. Kimple. 1996. Winter cover crops in a vegetable cropping system: Impacts on nitrate leaching, soil water, crop yield, pests, and management costs. Agriculture, Ecosystems, and Environment 59:1–17.

PHOTO: ISTOCKPHOTO.COM

CROP MANAGEMENT

5 Weed Control in Organic Strawberries

STEVEN A. FENNIMORE

Weed management is essential in organic strawberries because excessive weeds reduce yields and some weeds act as alternate hosts for soil pathogens. Weeds are harder to manage in organic strawberry operations because a number of weed management tools that are available to conventional growers, such as soil fumigation and herbicides, are not available to the organic grower (Daugovish and Fennimore 2008). The primary weed management tools for organic strawberry growers are field selection, sanitation, hand weeding, crop rotation, soil solarization, and plastic mulches.

SOIL SEED BANKS

In cropland, the term *soil seed bank* refers to the weed seed populations in the soil. Most annual weeds that infest a field come from the soil seed bank and are the direct result of weeds from previous seasons going to seed in the same field. Management of the soil seed bank affects the future severity of weed infestations in a particular field. In other words, successful weed management in rotational crops can reduce the effects of weeds in the subsequent strawberry crop. While it is not economically feasible to completely eliminate weed seed production, careful weed management can reduce the number of weed seeds per acre in the seed bank and decrease the future weediness of a field.

FIELD SELECTION AND SANITATION

Field selection is a critical first step in the organic strawberry weed management program because there are so few economical weed control options available. As a grower, you need to be aware of the level of weed infestation in a field before you can decide whether organic strawberries will be a viable crop there. It is essential that you avoid fields infested with perennial weeds like Bermudagrass, field bindweed, johnsongrass, or nutsedge because these weeds cannot be controlled economically in an organic production field. Fields infested with common purslane are hard to manage because the weed tolerates uprooting and can reroot and set seed in the field, further increasing the soil seed bank (Haar and Fennimore 2003). If common purslane is present, you need to uproot the plants and remove them from the field.

Annual weeds such as common groundsel, common sowthistle, hairy nightshade, and shepherd's-purse pose a problem for strawberry producers because they are potential hosts for Verticillium wilt. Growers can reduce the density of these weeds in the soil seed bank simply by growing an intensively cultivated rotation crop such as lettuce or a cole crop. Specific examples showing how to minimize the entry of weed seed into the field will be given later in this chapter.

It is important to control weeds in the vicinity of the field because windblown seed from common groundsel, common sowthistle, and hairy fleabane can enter the field's seed bank from surrounding areas such as fencerows and ditch banks. Weed control in the areas around the field is necessary to prevent contamination of the seed bank in the strawberry field. Similarly, it is essential that you clean any tillage equipment before moving it from fields infested with perennial weeds or other hard-to-control weeds and bringing it into uninfested fields. Use a pressure washer to clean the equipment of soil from the infested field, since that soil will carry with it weed seeds and other weed propagules.

CULTIVATION AND HAND WEEDING

Preplant irrigation to stimulate weed growth can improve the level of weed control before transplanting by depleting the bank of weed seeds from the upper soil layer (Shem-Tov, Fennimore, and Lanini 2006). After beds are formed, the field is irrigated to stimulate weed emergence. Approximately 7 to 14 days after preplant irrigation (longer in clay soils), you can remove emerged weeds with shallow tillage, organic-compliant herbicides, or propane flaming. This establishes a clean soil layer, depleted of weed seeds, close to the soil surface (figure 5.1). If you till the soil after this point, it is important that you keep the tillage implement on a shallow setting so the clean soil near the surface will not get contaminated with viable weed seeds that lie deeper in the soil. Many weed seeds will remain in deeper layers, but most of these deep seeds will not germinate. If time permits, you can repeat the process of preplant irrigation a second time to remove weeds that escaped the first preplant irrigation cycle.

The role of mechanical cultivation during the growing season is limited because of the plastic mulches used in organic strawberry production. Depending on how the plastic is placed, you may still be able to cultivate the furrow bottoms. In areas where grasses are planted in the furrow bottom, though, mechanical cultivation simply is not possible. The strawberry transplant will grow through a hole in the plastic mulch, and if you keep that hole as small as possible, you can keep weed growth through the hole to a minimum. After you install the plastic mulch, hand weeding is the only significant weed control option available to you. An important consideration in hand weeding is that you must remove weeds before they can set seed. Weed species such at burning nettle, common purslane, and shepherd's-purse can set seed within 21 to 35 days after emergence (Haar and Fennimore 2003; Fennimore and Roth 2003). Weed removal before seed set is essential to breaking the life cycle of weeds and keeping the field from becoming weedier.

CROP ROTATION FOR WEED MANAGEMENT

To keep weeds under control, growers should rotate organic strawberries with intensively cultivated crops such as celery, cole crops, and lettuce. Intensive cultivation and hoeing in these crops help remove most weeds and minimize weed seed production. Cole

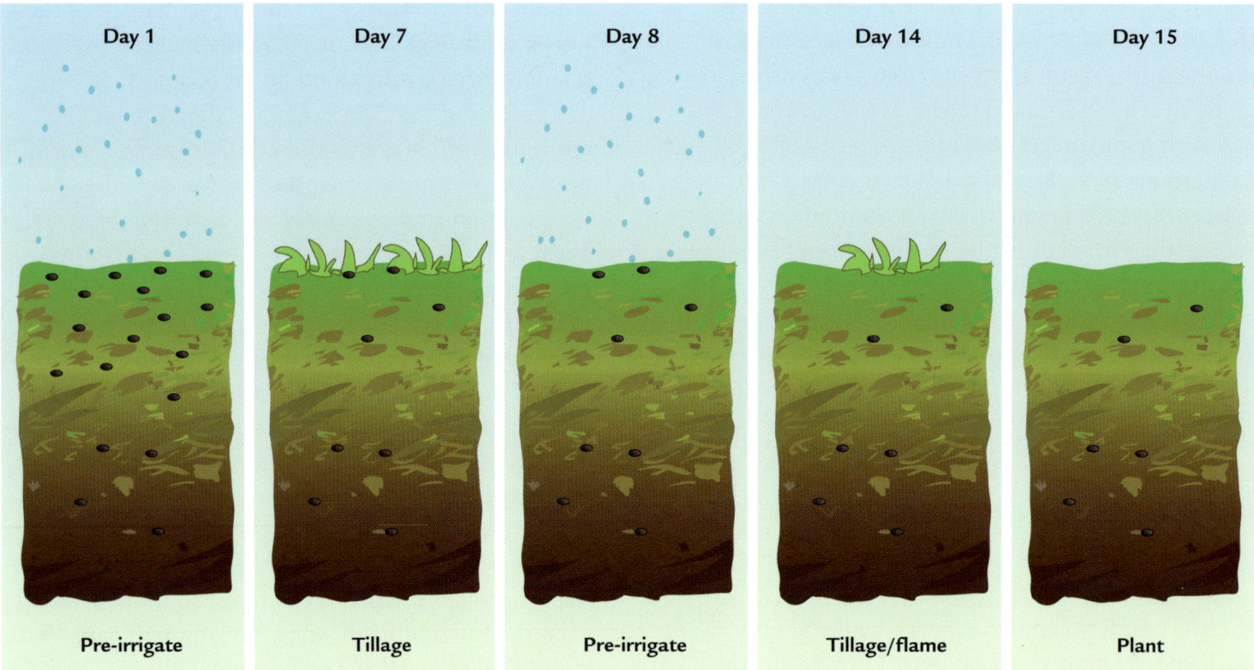

Figure 5.1. Weed removal using preplant irrigation followed by tillage or flaming. The schedule of events is preplant irrigation on days 1 and 8 followed by shallow tillage or flaming on days 7 and 14. With each round of irrigation, growth, and weeding, fewer weed seeds are left in the soil.

crops and mustard cover crops also have the advantage that they provide some level of weed control as a result of allelopathic compounds that they produce. *Allelopathy* is a term that refers to the "direct or indirect effects of one plant upon another through the production of chemical compounds released into the environment" (Rice 1984). In other words, allelopathic plants produce natural compounds that have herbicidal properties. When cover crops such as small grains, legumes, or mustards are planted singly or in combination, you need to make sure that planting densities and varietal selection are such that the cover crops will prevent weeds from setting seed (Brennan and Smith 2005).

The use of synthetic fumigants is not an option for organic strawberry producers. However, there is some interest in the use of mustard cover crops (*Brassica* spp.) and mustard meal as biofumigants, a form of allelopathy. Considerable research in the Pacific Northwest has assessed the feasibility of using mustard cover crops to suppress seeds and soilborne pathogens such as *Verticillium dahliae* in potatoes. Residues from a mustard cover crop can reduce the growth of weeds (Bialy et al. 1990). Glucosinolates in mustard cover crop residues are the major chemical components responsible for the fungicidal, herbicidal, and nematicidal activity of *Brassica* spp. (Fenwick, Heaney, and Mawson 1989). As the plant tissues break down in soil, glucosinolates are converted to isothiocyanates, thiocyanates, and other compounds that help with disease and weed control. Rapeseed foliage incorporated into the soil has been shown to control common lambsquarters and redroot pigweed to a degree nearly equal to that provided by a standard herbicide treatment (Boydston and Hang 1995). However, more modest control (i.e., 30 to 40%) of redroot pigweed and velvetleaf was observed in soybeans (Krishnan, Holshouser, and Nissen 1998). Recent results suggest that brassica seed meal does not control weeds, but it did increase strawberry fruit yield in relation to the untreated control (Fennimore et al. 2008). These results suggest potential benefits from planting mustard cover crops in organic strawberry fields, but further research and economic analysis are needed before we can accurately assess those benefits.

SOIL SOLARIZATION

Soil solarization can be an effective option in the Central Valley and other warm areas of California (figure 5.2) (Elmore et al. 1997). Most organic strawberries are grown in cooler locations along the coast, however, where results from soil solarization may be less satisfactory. In cooler areas, the heat from solarization does not penetrate very deep into the soil. However, because solarization kills most weed seeds in the upper layers of the soil, good weed control can still result (Gilbert et al. 2007, 2008). The use of double-layer tarps may improve the value of soil solarization in cool areas by increasing the soil temperature under the tarp. In double-layer applications, the bottom tarp is applied directly to the soil surface and a second layer is installed over a frame, leaving a 1- or 2-inch buffer space between the layers (Johnson 2003). The problem with the double-layer method is that no commercial product designed to perform this double-layer tarping procedure is available on the market. Another idea for improving results from solarization is to use it in combination with broccoli or other cole crop residues incorporated into the soil prior to solarization (Stapleton, Elmore, and DeVay 2000). This technique still requires further research.

MULCHES

A mulch can consist of plastic tarps, fiber mats, or straw and other organic material. Mulches serve multiple functions, such as warming or cooling the soil, protecting the fruit and foliage from the soil, enhancing moisture retention, and controlling weeds. Clear tarps are most useful for warming the soil and encouraging early development of the crop, but they can enhance weed growth in nonfumigated soils and are not recommended for organic systems. Synthetic mulches are available in black, blue, brown, green, red, white, and yellow. Some of the colors (including white and yellow) are available on a black background, which greatly improves weed control. Blue, red, and clear plastic mulches give the poorest weed control because they permit the greatest amount of light to penetrate the plastic, often enough to allow weed germination and growth under the plastic (Johnson and Fennimore 2005). Black, brown, and green mulches block light effectively and are recommended for use in organic systems. Growers should proceed with caution when using green tarps, since there is considerable variation in the amount of light intercepted by green mulches so levels of weed control can also vary.

Organic mulches such as wood chips, chopped straw, and rice hulls are sometimes used to control weeds. To control weeds effectively, this kind of mulch must be at least 2 inches deep. Organic mulches may increase problems with slugs, snails, earwigs, and other soil-dwelling pests. Researchers in Minnesota evaluated wool mat mulching for its ability to suppress weeds in strawberry fields (Forcella et al. 2001). Wool mats 0.4 inch thick were installed in the field at the time of transplanting, and weed control was evaluated for a period of 2 years after installation in perennial culture strawberries. The wool mats provided weed suppression for about one season before they decomposed.

ORGANIC-COMPLIANT HERBICIDES, FLAME WEED CONTROL

Organic-compliant herbicide options are limited and play only a minor role in weed management in organic strawberries. Potential organic-compliant herbicides include acetic acid (vinegar), fatty acids, soaps, and natural oils such as clove oil (Boyd and Brennan 2006). Bingaman and Christians (1995) suggest that corn gluten meal (CGM) may have herbicidal properties. Researchers evaluated CGM as a strawberry herbicide in Iowa (Bingaman and Christians 1995; Dilley, Nonnecke, and Christians 2002), where they found effective CGM rates to be approximately 2,900 pounds per acre, including 261 pounds of nitrogen. They found that CGM can provide a modest level of weed control, but that it may require repeat applications. Corn gluten meal is approximately 9 percent nitrogen, and there is some indication that its weed control activity is due to nitrogen toxicity to the weeds (Dilley, Nonnecke, and Christians 2002). However, evaluations in strawberries, turf, and vegetables in California have shown little or no weed control activity when CGM is applied alone (C. Elmore, unpublished results; S. Fennimore, unpublished results) (table 5.1). The Organic Materials Review Institute (OMRI) lists CGM as a restricted product, so it is unclear whether it can be used as an herbicide in organic strawberries.

Propane flaming is one of the more effective weed control tools available to organic producers (Boyd, Brennan, and Fennimore 2006). In flame weeding, you use a propane torch to remove weeds before you transplant strawberries into the field. Propane flaming is limited to pretransplant use because of the damage the flames will do to plastic mulch. Flame weeding does not burn weeds; rather, it heats weed tissue rapidly so that the plant cells rupture and die. The site of the weed's growing point determines the degree of success of flame weeding. Broadleaf weed seedlings are easier to kill with flaming than are grass weeds because the growing point of a broadleaf weed is exposed, whereas the growing point of a grass is more protected. It is impossible to control perennial weeds with a single propane application because only the exposed portions of the plant are killed. Perennial weeds such as Bermudagrass, johnsongrass, nutsedge, and field bindweed have large rhizomes or tubers underground that are beyond the reach of flaming. Perennial weeds quickly regrow after the aboveground foliage is killed. It is not economical to raise organic strawberries in a field infested with perennial weeds.

A number of products that contain acetic acid, clove oil, and lemon extract have potential as pre- or post-transplant herbicides in organic strawberries. Matran EC (clove oil) is registered as a directed spray in organic strawberries at 5 to 8 percent v/v (volume-to-volume) (CDMS 2008). Research reports indicate that acetic-acid–based products can provide effective weed control under some circumstances but that their performance is inconsistent (R. Smith and T. Miller, personal communication; Boyd, Brennan, and Fennimore 2006). More research is needed in this area.

ORGANIC VS. CONVENTIONAL CULTURE

Weed infestations are different in organic production fields than in conventional fields. For example, grass weeds are more predominant in organic strawberry fields. In the Central Coast area of California, annual bluegrass is the predominant grass weed. Bluegrass has an annual emergence cycle that peaks in November (Shem-Tov and Fennimore 2003). To deplete annual bluegrass seedbanks, we recommend careful management of this weed in rotation crops planted 1 or 2 years prior to the strawberry planting. For example, the fall before you are to plant a field to strawberries, you can plant it with an easy-to-weed crop such as cauliflower so you will be able to deplete the seed bank and keep it clean and ready for strawberries. Careful management of season-specific weeds such as annual bluegrass will help deplete the seed bank of those weed species.

RESEARCH NEEDS

One research need in this area is for a thorough characterization of the weed control efficacy of organic-compliant herbicides. We also need more and better compliant herbicides that can be used as directed sprays for row middles in organic strawberries.

A field index system that would allow growers to characterize the health of an organic production field would be of tremendous benefit. Such an index would take into consideration the health of the soil

Table 5.1. The influence of corn gluten meal (CGM) applications at 871 pounds per acre on weed densities in organic strawberries at Salinas, California

Treatment	Annual bluegrass	Corn spurry	Hairy nightshade	Shepherd's-purse
	density (number/2,652 cm^2)			
CGM	19.6	24.0	10.4	3.0
Untreated	17.9	23.0	15.0	2.6
LSD 0.05	ns	ns	ns	ns

Figure 5.2. Treatment of strawberry beds with solarization. Strawberries are planted following 4 to 6 weeks of soil solarization. *Photo* by S. Fennimore.

in terms of soil pests and could yield an overall score that could be used to determine whether the field should be considered for organic strawberry production. Specific aspects of the index would focus on the level of weed infestation and the level of soilborne disease infestation in the field.

Biodegradable mulches that provide effective weed control would also be a benefit. Plastic mulches are hard to recycle, and most are hauled to landfills at the end of the season.

Cover crop research could be done to better define the weed control benefits of planting various cover crops, such as small grains and mustards, prior to organic strawberry production.

REFERENCES/RESOURCES

Bialy, Z., W. Oleszek, J. Lewis, and G. R. Fenwick. 1990. Allelopathic potential of glucosinolates (mustard oil glucosides) and their degradation products against wheat. Plant and Soil 129:277–281.

Bingaman, B. R., and N. E. Christians. 1995. Greenhouse screening of corn gluten meal as a natural control product for broadleaf and grass weeds. HortScience 30:1256–1259.

Boyd, N. S., and E. B. Brennan. 2006. Burning nettle, common purslane, and rye response to a clove oil herbicide. Weed Technology 20:646–650.

Boyd, N. S., E. B. Brennan, and S. A. Fennimore. 2006. Stale seedbed techniques for organic vegetable production. Weed Technology 20:1052–1057.

Boydston, R. A., and A. Hang. 1995. Rapeseed *(Brassica napus)* green manure suppresses weeds in potato *(Solanum tuberosum)*. Weed Technology 9:669–675.

Brennan, E. B., and R. F. Smith. 2005. Winter cover crop growth and weed suppression on the Central Coast of California. Weed Technology 19:1017–1024.

CDMS (Crop Data Management Systems). 2008. Matran EC sample label. Accessed at www.cdms.net/LDat/ld77P003.pdf.

Daugovish, O., and S. A. Fennimore. 2008. Weeds. Pp. 115–130 in M. L. Flint (ed.), UC Integrated Pest Management for Strawberries, Second Edition. Oakland: University of California Division of Agriculture and Natural Resouces. Publication 3351.

Dilley, C. A., G. R. Nonnecke, and N. E. Christians. 2002. Corn-based extracts to manage weeds and provide nitrogen in matted-row strawberry culture. HortScience 37:1053–1056.

Elmore, C. L., J. J. Stapleton, C. E. Bell, and J. E. DeVay. 1997. Soil solarization: A nonpesticidal method for controlling diseases, nematodes, and weeds. Oakland: University of California Division of Agriculture and Natural Resources. Publication 21377.

Fennimore, S. A., H. Ajwa, K. Subbarao, F. Martin, G. Browne, and S. Klose. 2008. Facilitating adoption of alternatives to methyl bromide in California strawberries. In Annual International Research Conference on Methyl Bromide Alternatives and Emissions Reductions. Orlando, Florida. Proceedings 11.

Fennimore, S. A., and K. A. Roth. 2003. Predicting time to emergence, flowering, and seed set in burning nettle *(Urtica urens)*. Weed Science Society of America Abstracts 43:23–24.

Fenwick, G. R., R. K. Heaney, and R. Mawson. 1989. Glucosinolates. Pp. 97–141 in P. R. Cheeke (ed.), Toxicants of plant origin. Volume II: Glucosides. Boca Raton: CRC Press.

Forcella, F., S. Poppe, N. Hansen, and E. Hoover. 2001. Weed management with bio-based mulches in transplanted strawberry. Weed Science Society of America Abstracts 41:106–107.

Gilbert, C. A., S. A. Fennimore, K. Subbarao, R. Goodhue, and J. B. Weber. 2007. Solarization and steam heat for soil disinfestation in flower and strawberry. In Annual International Research Conference on Methyl Bromide Alternatives and Emissions Reductions. San Diego, California. Proceedings 109.

Gilbert, C. A., S. A. Fennimore, K. Subbarao, R. Goodhue, J. B. Weber, and J. Samtani. 2008. Soil disinfestation with steam and solarization for flower and strawberry. In Annual International Research Conference on Methyl Bromide Alternatives and Emissions Reductions. Orlando, Florida. Proceedings 68.

Haar, M. J., and S. A. Fennimore. 2003. Evaluation of integrated practices for common purslane management in lettuce. Weed Technology 17:229–233.

Johnson, M. 2003. Nonchemical weed control inputs for strawberry. MS Thesis. University of California, Davis.

Johnson, M. S., and S. A. Fennimore. 2005. Weed and crop response to colored plastic mulches in strawberry production. HortScience 40:1371–1375.

Krishnan, G., D. L. Holshouser, and S. Nissen. 1998. Weed control in soybean *(Glycine max)* with green manure crops. Weed Technology 12:97–102.

Rice, E. L. 1984. Allelopathy, Second Edition. New York: Academic Press.

Shem-Tov, S., and S. Fennimore. 2003. Seasonal changes in annual bluegrass (*Poa annua* L.) germinability and emergence in California. Weed Science 51:690–695.

Shem-Tov, S., S. A. Fennimore, and W. T. Lanini. 2006. Weed management in lettuce *(Lactuca sativa)* with pre-plant irrigation. Weed Technology 20:1058–1065.

Stapleton, J. J., C. L. Elmore, and J. E. DeVay. 2000. Solarization and biofumigation help disinfest soil. California Agriculture Journal 54(6):42–45.

PHOTO: KATHY KEATLEY GARVEY

CROP MANAGEMENT

6

Nitrogen Management for Organic Strawberries

JOJI MURAMOTO AND MARK GASKELL

Nitrogen (N) is a key nutrient for growing strawberries. The plant's roots absorb it from the soil, mainly in the form of inorganic nitrogen: nitrate (NO_3^-) or ammonium (NH_4^+). (Nitrogen can be divided into organic [e.g., amino acids, protein] and inorganic forms. In this chapter, *inorganic nitrogen* refers to nonorganic forms of nitrogen such as ammonium and nitrate.) The plant then assimilates absorbed inorganic nitrogen into amino acids and proteins that it uses to build new cells. An adequate supply of nitrogen is necessary for the development and growth of shoots, leaves, flowers, and fruit.

Organic growers use organic sources of nitrogen or nitrogen derived from natural processes to add and maintain soil organic matter (SOM). To be used in an organic field, fertilizers have to be certified by the Organic Materials Review Institute (OMRI) or another certifier. Contact a local organic certifier to confirm that the material is allowed for use in organic fields in your area (see chapter 12, *Organic Certification and Registration in California,* for more information). A major portion of the nitrogen in these sources and in SOM exists as organic nitrogen. These organic nitrogen materials have to undergo microbial decomposition (mineralization) before they become available for plant uptake. Many factors affect the rate of mineralization, and this makes the timing of nitrogen availability in organic systems unpredictable.

The goal of nitrogen management in organic strawberry production is to synchronize the nitrogen supply from organic fertilizers and SOM with the nitrogen demand of the strawberries. On the one hand, a lack of nitrogen at critical times can limit plant development and, eventually, fruit production, while on the other hand excessive nitrogen applications can cause undesirable nitrogen losses into the environment. Nitrogen management in organic strawberry production is challenging for many reasons. In this chapter, we will start out by reviewing general aspects of nitrogen nutrition in strawberries. Then we will go on to discuss specific nitrogen management issues that come up in organic strawberry production.

EFFECT OF NITROGEN ON STRAWBERRIES

The yield and quality of strawberry fruit is strongly influenced by the status of nitrogen in the plant. Furthermore, the plant's nitrogen demand changes with each growth stage and is affected by differences in growing region, cultivar, year, cultural practices, and the quality of the transplants (i.e., the amount of cumulative chill hours and handling conditions after the transplants are harvested in the nursery).

A lack of nitrogen during early vegetative growth (between transplanting and flowering) limits the development of crown axillary branches in number and size, and that leads to lower fruit yields. There is a greater potential for this sort of deficiency in California's Southern Coast strawberry production region (Ventura and San Diego Counties), where the temperature after transplanting is warmer and plants grow more quickly. Under adequate soil water management, the nitrogen need of strawberries early in the season is determined by temperature, with warmer, more southerly districts having greater nitrogen demands early on and a continuing need for nitrogen extending out weeks and months, in contrast to lesser demands in cooler, more northerly growing regions (figure 6.1) (Gaskell et al. 2009). During the reproductive growth period (from flowering to the end of harvest), nitrogen deficiency can reduce flower quality, leaf area, root mass, and fruit size, resulting in lower fruit yield and quality. These symptoms are often observed in California organic strawberries if the amount of supplemental nitrogen is insufficient during this stage.

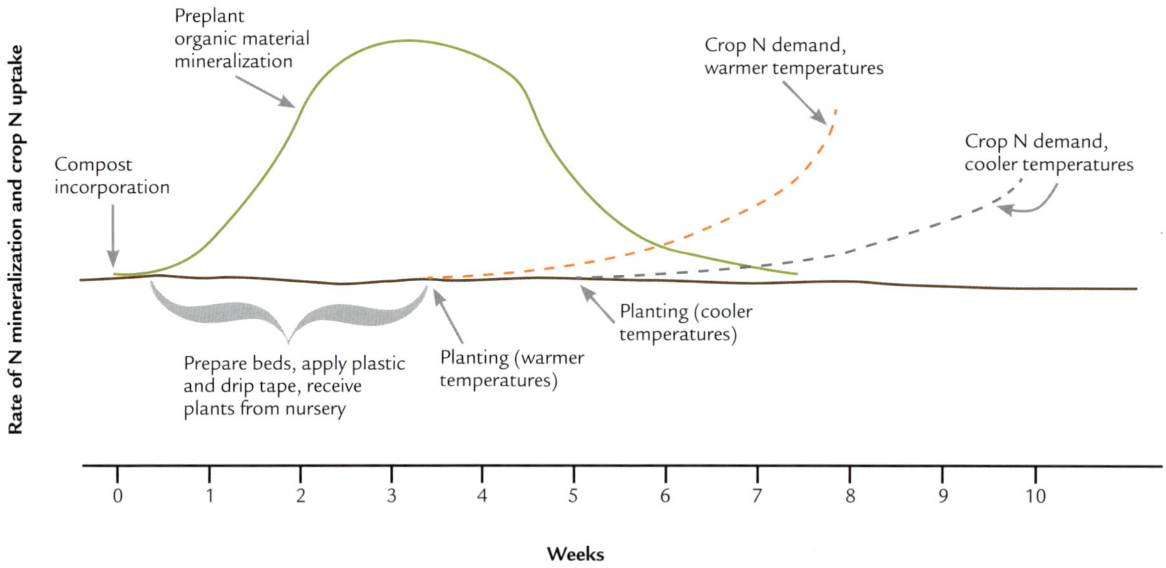

Figure 6.1. Graphic summary of relationship between preplant applied organic N mineralization, regional difference in anticipated strawberry N demand, and resulting asynchrony between preplant N supplies and strawberry N demand (Gaskell et al. 2009).

Conversely, excessive nitrogen at an early stage can delay reproductive bud differentiation, promote excessive vegetative growth, and reduce fruit size. Excessive levels of nitrogen at later stages can also lead to delayed ripening, soft fruit that are more susceptible to molds and powdery mildew, and lower yields (May and Pritts 1990). These problems can be found in conventional strawberry production in California, but it is rare to find excessive nitrogen conditions in organic strawberry production in the state.

NITROGEN UPTAKE BY STRAWBERRIES: PATTERN AND AMOUNT

Few studies are available on the uptake of nitrogen by organic strawberries. An example of the daily cumulative nitrogen uptake pattern for winter-planted organic strawberries (Aromas) grown in Monterey County is shown in figure 6.2 (Muramoto, unpublished data). In this example, the average marketable fruit yield was 37,700 pounds per acre, with a total yield of 54,600 pounds per acre. In the early growth stage (the first 15 weeks after transplanting), the plants slowly accumulated nitrogen under cooler climate conditions. When fruit harvest began in April or May, 4 to 5 months after transplanting, a steep increase in nitrogen uptake began. This appears to be a typical pattern for cumulative nitrogen uptake by a day-neutral strawberry cultivar grown in a winter-planted organic system in the northern strawberry production region of California (Monterey and Santa Cruz Counties).

In reality, however, there are many factors—including growing region, production cycle, cultivar, and climatic conditions during the specific year—that influence the length of each growth stage and the actual nitrogen demand of strawberry plants. For instance, Southern California's warmer temperatures make the period between transplanting and fruit harvest much shorter, with harvest for an early October–transplanted field starting in mid December, only 5 to 6 weeks after the transplanting date. Plants there develop rapidly so their nitrogen demand is greater in early stages than it is further north (figure 6.1).

Still, even with the long growth period, the amount of nitrogen uptake for strawberries is relatively low when compared to that of other shorter-cycle vegetable crops. In the example illustrated in figure 6.2, total nitrogen uptake for organic strawberries was close to 100 pounds per acre in 46 weeks (note that total nitrogen uptake amount can vary depending on many factors,

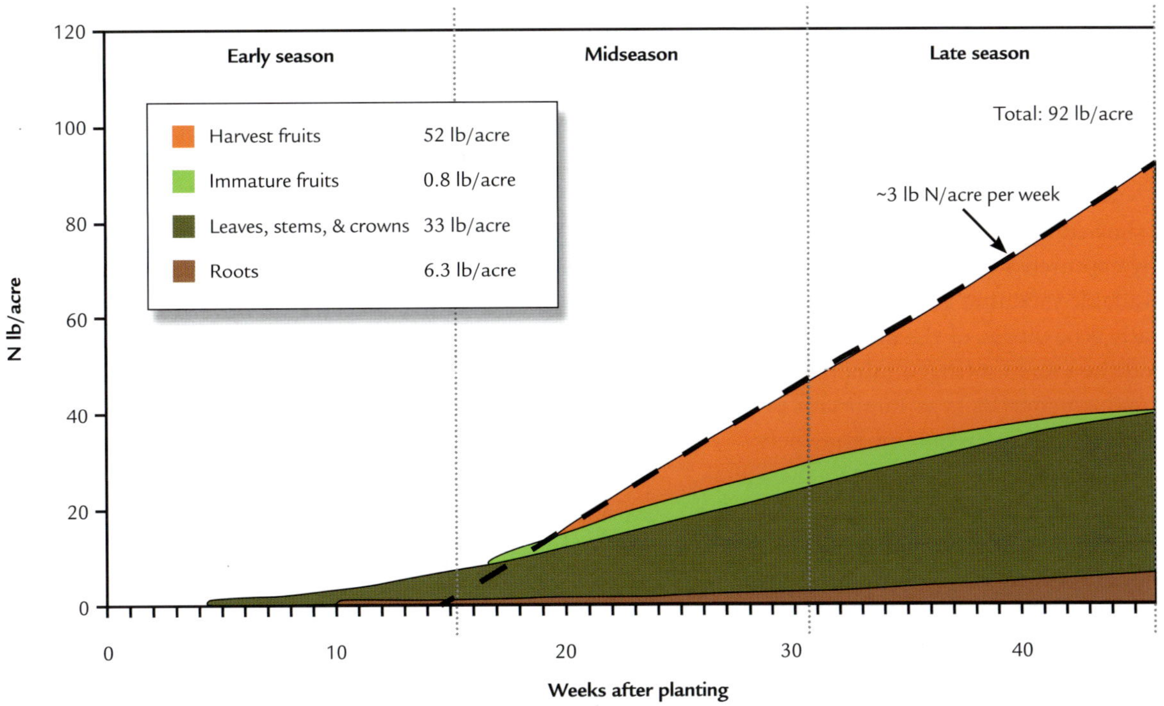

Figure 6.2. An example of cumulative N uptake and partitioning of organic strawberries on the Central Coast of California. (Cultivar: Aromas. Based on average data of four seasons from 2002 to 2006. Average fruit yield: marketable 37,700 lb/acre, total 54,600 lb/acre. Average fruit moisture: 91%. Average fruit N: 0.98% dry weight. Planting density: 20,580 per acre. Field location: Moss Landing, Monterey County, California. Soil type: Santa Ynez sandy loam (fine, montmorillonitic thermic, Ultic Palexerolls). Planting dates were in November or December of each year.

including plant population, yield, and nitrogen application rate). This is less than half the nitrogen uptake for a vegetable crop such as broccoli, which absorbs more than 200 pounds of nitrogen per acre during a growth period of approximately 15 weeks and reaches daily uptake levels of 5 to 7 pounds of nitrogen per acre (Sullivan, Hart, and Christensen 1999). In contrast, the strawberries described in figure 6.2 had a nitrogen uptake rate during the mid to late season of approximately 3 pounds of nitrogen per acre per week, suggesting that a low but steady supply of nitrogen is crucial during the long harvest period. More studies are needed to give us a fuller understanding of the effects of production region, cultivar, year, fruit yield, and nitrogen fertility management practices on the amount and rate of nitrogen uptake by strawberries.

NITROGEN MANAGEMENT WITH ORGANIC FERTILIZERS

Organic strawberry growers apply various types of organic materials to "feed the soil" and to meet the nitrogen demand of strawberry plants. Most of the nitrogen in these materials is in the form of organic nitrogen, and it must undergo a decomposition process (mineralization) before it can become available for plant uptake. Soils vary in organic matter content however, and many traditional strawberry production districts in California may have soils with relatively low organic matter content (i.e., 1% or less). Previous studies in California have shown that it is difficult to increase SOM by even as little as 1 percent, so the nitrogen from SOM has to be supplemented with additional nitrogen applications if a grower wants to get optimum strawberry yield and quality.

MINERALIZATION

Mineralization is a general term for the microbial decomposition processes that change organic materials into inorganic (mineral) compounds. In most warm, well-aerated soils, microbes metabolize organic carbon (C) in the organic materials, allowing the organic nitrogen to mineralize to NH_4^+ (ammonium), after which a subsequent microbial process (nitrification) quickly oxidizes the NH_4^+ to NO_3^- (nitrate). Some organic-nitrogen compounds (e.g., amino acids) can be acquired by plant roots, but the quantitative importance of this kind of uptake is debatable (Nashölm, Kielland, and Ganeteg 2009). One study showed that the strawberry cultivar Tribute only captured a small portion of nitrogen that was available in amino acid form (Reeve et al. 2008). Clearly, knowledge of the amount of applied nitrogen and the rate of nitrogen mineralization from the organic materials is as important to nitrogen management in an organic strawberry system as it is in other organic production systems (Mikkelsen and Hartz 2008).

As a biological process, the mineralization rate is affected by environmental conditions such as soil temperature (it is active above 50°F) and soil moisture (it proceeds quickly in moist soil and slows down in dry or poorly drained soil). Management practices such as tillage (mineralization is temporarily increased by soil tillage) and the makeup of the organic materials (C:N ratio and lignin content) also affect mineralization.

C:N RATIO, LIGNIN CONTENT, AND NITROGEN AVAILABILITY

The ratio of carbon to nitrogen content in the organic material, usually referred to as the C:N ratio, influences the rate of mineralization. The C:N ratio of organic materials may vary widely, from 3:1 to 4:1 for organic fertilizers such as blood meal and feather meal, to 15:1 for young plants, to between 50:1 and 80:1 for the mature straw of crops, to over 100:1 for sawdust. Usually the C:N ratio of soil is between 10:1 and 12:1; for soil microbes it is about 7:1. Even a material like compost, which is widely used by organic growers, can vary in C:N ratio from 12:1 or 15:1 up to 30:1 or more depending on the composting materials and the process used.

Microbes need both carbon and nitrogen for growth and reproduction. If a material has a C:N ratio greater than about 25:1, microbe populations will be limited by the small amount of nitrogen, the rate of decomposition will be slow, and NO_3^- and NH_4^+ in the soil will be tied up by the microbes. This condition is referred as immobilization of nitrogen, and it causes nitrogen deficiency in crop plants. Take care when applying such organic materials to the soil.

In contrast, if the material has a C:N ratio of 20:1 or 25:1 or less, the microbes are carbon limited and the rate of decomposition will be fast, leading to the release of excess nitrogen into the soil as NH_4^+ (Chaney, Drinkwater, and Pettygrove 1992). The decomposition rate for organic materials is also influenced by their lignin content. Materials high in lignin (e.g., bark, wood chips, corn stalks, etc.) decompose slowly even in the presence of excess nitrogen because lignin's chemical bonds are resistant to decomposition.

SOIL ORGANIC MATTER

Soil organic matter plays an important role in many aspects of organic production, including nitrogen management. Although it is generally present in small amounts (0.2 to 2% in most agricultural soils in California), SOM has multiple effects on the overall physical, chemical, and biological properties of a soil.

The SOM consists of three parts: living organisms, an active component (fresh and more readily decomposable residues), and a passive component (well-decomposed and more resistant residues) (Magdoff and van Es 2000). As far as nitrogen management is concerned, the living organisms component of SOM helps decompose organic materials and mix them with the soil minerals, thus recycling nitrogen. The decomposition of organic matter from plants and animals by these living organisms leads to formation of humic substances.

The passive component consists in large part of humus, which is more resistant to decomposition and may in fact last for hundreds of years. High C:N ratio crop residues such as straw are most often converted to humus. The increase of humus levels in soils is important because it provides a long-term source of nutrients and also because it improves soil characteristics, both chemical (increased cation-exchange capacity) and physical (stabilized aggregates and improvements in water infiltration and water-holding capacity).

The remaining fraction of SOM is made up of recently deceased organisms, plant litter, nonlignin crop residues, and other unrecognizable organic residues. This fraction of SOM is referred to variously as the active component, particular organic matter, or the light fraction. The active component is labile and readily decomposed. It contributes more directly to nutrient dynamics than do the other SOM components. Low C:N ratio crop residues such as vegetable crop residues produce less humus, decompose more quickly, and contribute more to the light fraction than do high C:N materials. The active component affects nutrient cycles and plant nutrient availability in the short term (Gaskell and Smith 2007).

PREPLANT FERTILIZATION

Crop Residues and Cover Crops

When you grow strawberries following a vegetable crop, you should take the nitrogen contribution from the vegetable residues into account as you determine the size of your preplant nitrogen application. For example, broccoli and lettuce crop residues respectively contain around 150 and 60 pounds of nitrogen per acre (Mitchell et al. 1999).

Many organic strawberry growers use cover crops as part of their crop rotation with strawberries. Various types of winter cover crop (e.g., legumes, cereals, or mustards, or mixes of these) and summer cover crops (e.g., Sudangrass) can be grown in strawberry production regions in California. Biomass from these cover crops normally contains over 100 pounds of nitrogen per acre, though the rate of mineralization after incorporation varies depending on the C:N ratio of the material.

With their relatively high nitrogen content and their C:N ratio (<20), vegetable residues and leguminous cover crops decompose relatively quickly once they are incorporated into the soil in the summer-to-fall period. Most of the mineralization process for such cover crops occurs over the first 3 to 6 weeks after incorporation, after which the soil NO_3^- level will return to pre-incorporation levels by weeks 6 to 8 in moist, pH-neutral soils (Gaskell and Smith 2007). This means that the nitrogen supply from vegetable crop residues and cover crops usually is not adequate to fulfill the mid- to late-season nitrogen demand of strawberries, meaning that additional nitrogen applications will be necessary. See chapter 4, *Cover Crops for Nutrient Management in Organic Strawberry Production on California's Central Coast,* for more information on the use of cover crops in organic strawberry production.

Animal Manures

The use of raw animal manures in organic strawberry production in California may be limited. USDA's National Organic Program (NOP) restricts the use of raw animal manure in crop fields unless (1) it is applied to land used for a crop not intended for human consumption, (2) it is incorporated into the soil no fewer than 120 days prior to the harvest of a product whose edible portion has been in direct contact with the soil surface or soil particles, or (3) it is incorporated into the soil no fewer than 90 days before the harvest of a product whose edible portion does not have direct contact with the soil surface or soil particles (USDA NOP 2011). Under such restrictions, limited numbers of organic strawberry growers may be able to use raw animal manures. The lack of uniform nitrogen content for these materials and their potential for contamination with human pathogens are also major issues (Mikkelsen and Hartz 2008; Chaney, Drinkwater, and Pettygrove 1992). Consult with your organic certifier before you apply raw animal manures to your organic fields.

Compost

Compost is a stable, relatively uniform organic amendment produced through aerobic microbial decomposition of various organic materials such as green wastes, crop residues, and animal manures. While a given lot of compost tends to be reasonably uniform and consistent, nitrogen and organic matter content of individual batches of compost may vary if they are made from different source materials, or even if they are made from the same source materials at different times of the year. The USDA National Organic Standard has regulations on composting processes as follows: you must (1) establish an initial C:N ratio of between 25:1 and 40:1 and (2) maintain a temperature of between 131° and 170°F for 3 days using an in-vessel or static aerated pile system or (3) maintain a temperature of between 131° and 170°F for 15 days using a windrow composting system, during which period the materials must be turned a minimum of five times (USDA NOP 2011).

Compost applications can increase soil organic matter content, improve the soil's biological diversity and activity, improve its overall quality and tilth, and increase its mineralizable organic nitrogen content. Compost is usually broadcast onto the field immediately after the cover crop or the preceding vegetable crop is incorporated and before the field is leveled and bedded. Application rates of 3 to 10 tons of compost per acre are common for organic strawberry growers in California.

Like other nutrients, nitrogen levels in composts vary greatly, but they generally range from 1 to 2 percent of dry weight. Most nitrogen in composts is in an organic form. Although it is difficult to predict what the rate of mineralization in the field will be over a period of years, mineralization rates for organic nitrogen in composts are low. Sample compositions and nitrogen mineralization rates for composts and manures as determined in a greenhouse study are shown in table 6.1. A different study (Hartz, Mitchell, and Giannini 2000) reported much lower mineralization rates for manures and composts from California than appear in table 6.1, with average mineralization rates for organic nitrogen in manure, manure compost, and plant residue compost after 24 weeks' incubation of 15, 6, and 2 percent, respectively. Based on this information, it is unlikely that the nitrogen demands of a strawberry field can be satisfied by compost alone.

A larger amount of annual compost application (i.e., applications in excess of 20 ton/acre per year) over multiple years is not recommended since it can eventually cause nutrient imbalances in the soil. For example, repeated applications of manure-based compost at a higher rate might cause excessive amounts of phosphorus to accumulate in the soil, since manure composts, and especially chicken manure composts, are relatively high in phosphorus. Excessive phosphorus in the soil will eventually move off the site, often attached to soil that erodes, and this may lead to the eutrophication of nearby surface waters. When soil tests indicate that you have high levels of available phosphorus in your soil, you should either reduce your compost application rate or choose compost with a low phosphorus content.

High-Nitrogen-Content Organic Fertilizers

In addition to compost, many organic strawberry growers apply commercial solid organic fertilizers such as blood meal, bonemeal, feather meal, or fish meal, or a pelletized mixture of these. These fertilizers are usually applied to the beds prior to planting by means of a fertilizer drill that works with the bed shaper. These organic fertilizers have a relatively fast mineralization rate. Studies have shown the duration of nitrogen release from such fertilizers to be 5 to 8 weeks (Hadas and Kautsky 1994; Hartz and Johnstone 2006). In the northern and central strawberry production regions of California, application of high-nitrogen-content organic fertilizers can boost soil inorganic nitrogen content well before the nitrogen demand peak of strawberries (figure 6.1). This asynchrony can cause undesirable nitrogen losses to the environment, as most soil nitrate at the time of planting, whether from the organic fertilizers or residual nitrate from the preceding crop, can be immobilized or lost to the environment during the winter rainy season under a Mediterranean climate (Muramoto et al. 2004).

SUPPLEMENTAL FERTILIZATION

Supplemental in-season applications of solid organic fertilizers are difficult if you use an annual plasticulture system with drip irrigation, as is currently typical in California fields. Instead, the application of liquid organic fertilizers through a drip irrigation system (fertigation) is gaining popularity among organic strawberry growers. Organic liquid fertilizers can include liquefied fish residues or fish emulsion, fish powder, blood meal, bat guano, feather meal, seabird guano, and composted manure teas. Liquid organic fertilizers are primarily made up of fine-ground suspended organic particles along with some soluble nitrogen. Some partially decomposed nitrogen may also be available from hot or cold hydrolysis of proteins.

The advantages of these fertilizers are that the nitrogen is more readily available to the plants and that there are virtually no restrictions on application timing. However, special problems associated with these materials limit their efficient use. First, the material's label commonly indicates a nitrogen content of 6 to 8 percent, but the amount of nitrogen that actually arrives at the root zone is less certain (Gaskell and Smith 2007). Second, uniformity is once again a problem because the fertilizers are suspensions of organic particles, and uniform, consistent application requires that they be frequently mixed (Gaskell et al. 2009). Further, careful management is required when you use fertigation, since there is a high potential for leaching nitrogen into the environment if the crop is overirrigated or if you practice leaching for soil salinity management. Foliar application can also be used to supplement nitrogen, but the effectiveness and

Table 6.1. Available nitrogen (N) from animal manures and composts

Manure type	Total N (% of dry wt.)	Inorganic N* (% of dry wt.)	C:N ratio	Available N in 10 months	
				% of total N	lb/ton dry wt.
Beef	2.6	0.02	14:1	40	21
Chicken	4.6	0.15	6.5:1	65	60
Chicken (composted)	1.7	0.22	6.5:1	35	12
Dairy-1	2.9	0.08	12:1	30	17
Dairy-2	2.0	0.04	16:1	24	10
Dairy (composted)-1	2.0	0.03	12:1	21	8
Dairy (composted)-2	1.9	0.03	10:1	14	5
Dairy (composted)-3	2.1	0.06	6.7:1	6	2
Pig	3.9	0.14	9.9:1	47	37
Sheep	2.3	0.08	13:1	36	17

* NO_3-N + NH_4-N

Source: Castellanos and Pratt 1981.

efficiency of the practice has not been well evaluated in organic strawberry culture.

Figure 6.3 shows typical nitrogen management strategies for organic strawberry production in different regions in California (Gaskell et al. 2009). As a grower, you can use this figure as a guideline to help you optimize your nitrogen management strategy for each field situation.

ADDITIONAL FACTORS

The cost per unit of nitrogen for commercial organic fertilizers is high, and your real cost may end up being higher still because of inefficiencies and loss of material during fertigation. Table 6.2 shows a list of organic fertilizers and their estimated cost per unit of nitrogen.

With their shallow root system, strawberries can also be a salinity- and water-stress–sensitive crop. As much as 50 to 90 percent of a strawberry plant's root system is in the top 6 inches of soil (UC IPM 2008). In addition, fine roots tend to develop below the drip tapes that supply water and nutrients (including nitrogen) during the long growing season (figure 6.4). Besides the contribution it makes to nitrogen management, efficient irrigation management is important for organic strawberry production because it helps growers to optimize the use of expensive inputs and eliminate adverse environmental impacts.

TOOLS FOR NITROGEN MANAGEMENT

Pre-Plant Soil Mineral Nitrogen Test

If you plant strawberries following a heavily

Preplant Soil Inorganic Nitrogen Test

If you plant strawberries following a heavily fertilized vegetable crop, the field may still have a significant amount of residual inorganic nitrogen in the soil. To avoid overfertilization, test the residual inorganic nitrogen content (nitrate N [NO_3-N] + ammonium N [NH_4-N]) in the soil before you make any preplant compost or organic nitrogen fertilizer applications.

Soils laboratories usually express soil inorganic N results in terms of parts per million (ppm). You can use the method shown below to convert lab results from ppm to pounds per acre in 6-inch-deep soil, but you need to know the soil's bulk density. For a fairly easy technique for measuring soil bulk density, refer to Soil Quality Institute (1998) in the References/Resources at the end of this chapter or check with a local UC Cooperative Extension specialist.

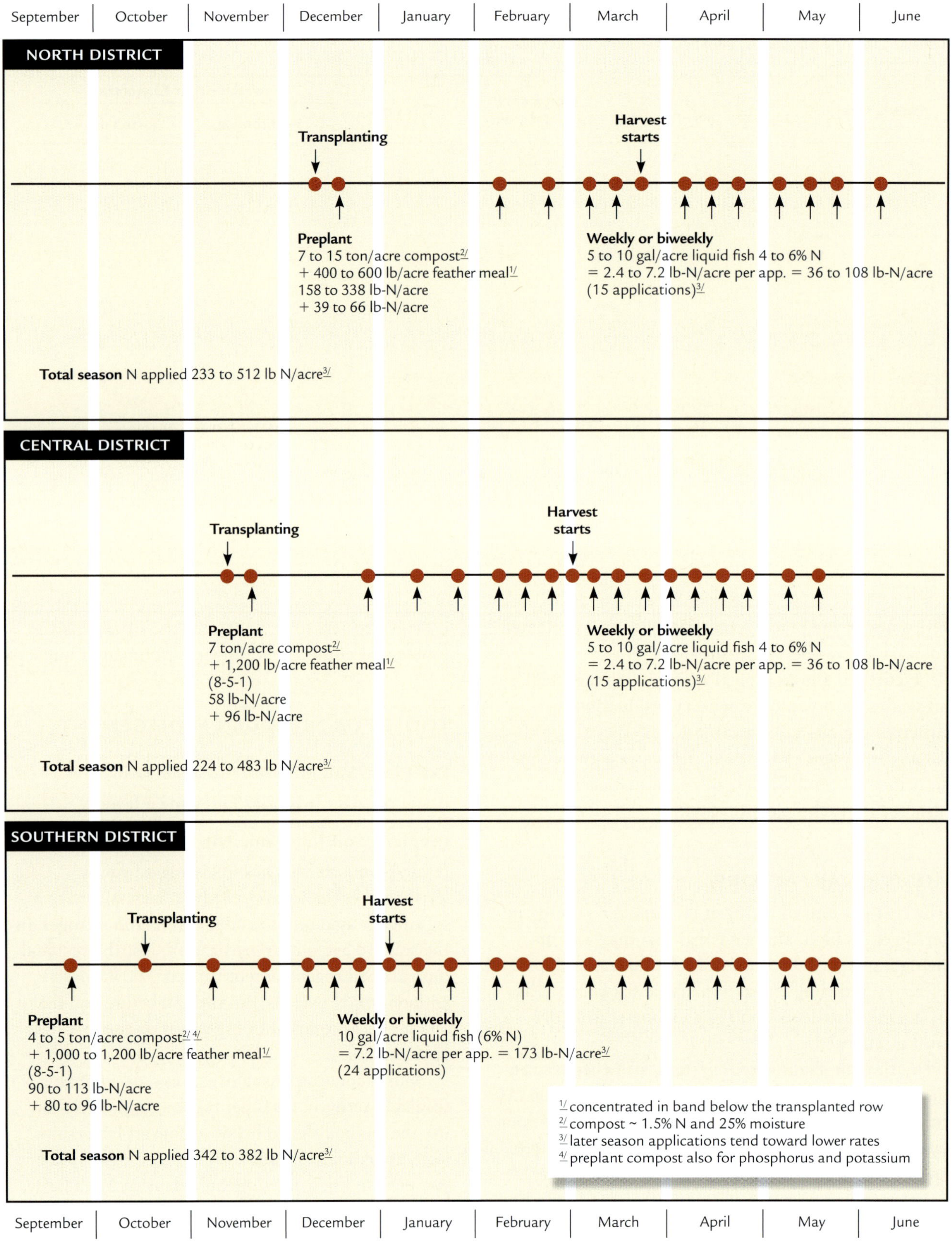

Figure 6.3. Current reported typical N fertilization programs for organic strawberry production in California (Gaskell et al. 2009).

Table 6.2. Total nitrogen (N) availability (initial inorganic N + mineralized organic N) and cost per pound of available N

Organic fertilizer	Total N availability		$/ton	$/lb available N
	% of initial N	lb/ton		
Blood meal	70	221	1,000	4.50
Feather meal	63	179	600	3.60
Fish powder	65	178	4,000	22.50
Poultry manure (pelleted)	46	26	70	2.70
Seabird guano	79	175	400	2.30
Seabird guano (pelleted)	74	173	700	4.00

Source: Hartz and Johnstone 2004.

Figure 6.4. An example of the vertical root distribution pattern of an organic strawberry plant, shown by means of the pin-board method. Nail grid spacing is 2 inches. The black circle indicates the location of the drip tape. Cultivar: Aromas; location: Moss Landing, Monterey County, California; soil type: Santa Ynez fine sandy loam (fine, montmorillonitic thermic, Ultic Palexerolls). *Photo* by J. Muramoto.

If your soil samples came from between 0 and 12 inches deep, use a soil depth factor of 3 rather than 1.5. In that case, for a soil with a bulk density of 1.5, soil inorganic N (lb/ac in 12-inch-deep soil) = inorganic N (ppm) × 4.

Adjust your preplant nitrogen application rate by taking the per-acre soil inorganic nitrogen content you have calculated and subtracting that from the rate you intended to apply to meet crop requirements.

Convert laboratory results (ppm) to field application levels (lb/acre):

inorganic N (ppm) × soil bulk density (1.0 to 1.6) × soil depth factor (1.5 for 6-inch-deep soil) × metric-to-English conversion factor (0.89)	=	soil inorganic N (lb/acre in 6-inch-deep soil)

For a 6-inch-deep soil with a density of 1.5:

inorganic N (ppm) × 1.5 × 1.5 × 0.89	=	soil inorganic N (lb/acre in 6-inch-deep soil)

or

inorganic N (ppm) × 2	=	1 lb/acre soil inorganic N in 6-inch-deep soil

what portion that is depends in part upon the growing area. For each 1 percent of organic matter in the soil, approximately 0.5 to 1 pound of nitrogen per acre may become available per day at coastal California sites during the strawberry production season. To give their fields good drainage, many growers plant strawberries on very coarse-textured soils with little organic matter content, so the importance of the nitrogen contribution from SOM can vary markedly from site to site. Consider the contribution from SOM as you calculate the overall nitrogen budget for your crop.

Plant Tissue Nitrogen Test

A plant tissue test is a useful tool for measuring the nitrogen status of a crop, and it has been used widely in conventional strawberry production (Maynard and Hochmuth 1997; Pritts and Handley 1998; Ulrich, Mostafa, and Allen 1980). In California, however, no field trial–based tissue test standards have been developed for strawberries, even for those farmed in conventional growing systems (UC IPM 2008). A recent study suggested that a tentative threshold value of 2,500 ppm NO_3-N in leaf petioles (UC IPM 2008) may be valid in the early flowering stage (April) of organically grown Seascape cultivar strawberries in a Northern California production region (Muramoto et al., unpublished data). Clearly, more studies are needed to establish a reliable tissue testing system for strawberries in California.

CONCLUSIONS

Nitrogen management in organic strawberries is a challenging area due to the importance of nitrogen to strawberry nutrition and the crop's long growth period. The unpredictability of soil nitrogen mineralization from organic fertilizers and SOM and the effect of temperature on strawberry growth and plant nitrogen demands make the application of preplant fertilizers prior to the rainy season in winter planting systems hard to calculate. Various differences between strawberry production regions also significantly affect nitrogen management strategies for organic strawberries. Some asynchrony in organic strawberry production is unavoidable, just as it is in other annual crops (Crews and Peoples 2005). A combination of limited preplant nitrogen (from compost and/or cover crop incorporation and lower levels of high-nitrogen-content organic fertilizer applications) combined with supplemental nitrogen applications using liquid nitrogen organic fertilizers applied via fertigation seems to be a reasonable approach to synchronizing the nitrogen supply with the nitrogen uptake needs of organic strawberries.

More research is needed to demonstrate differences in nitrogen uptake patterns and rates among different strawberry cultivars and growing environments, to examine ways to better synchronize nitrogen mineralization from organic fertilizers with the nitrogen demand of strawberries in different growing environments and to develop a reliable nitrogen tissue test system for strawberries in California.

ACKNOWLEDGMENTS

The authors greatly appreciate the valuable comments we received from two anonymous reviewers and from Mark Bolda on a previous version of the manuscript for this chapter.

REFERENCES/RESOURCES

Castellanos, J. Z., and P. F. Pratt. 1981. Mineralization of manure N correlation with laboratory indexes. Soil Science Society of America Journal 45:354–357.

Chaney, D. E., L. E. Drinkwater, and G. S. Pettygrove. 1992. Organic soil amendments and fertilizers. Oakland: University of California Division of Agriculture and Natural Resources. Publication 21505.

Crews, T. E., and M. B. Peoples. 2005. Can the synchrony of nitrogen supply and crop demand be improved in legume and fertilizer-based agroecosystems? A review. Nutrient Cycling and Agroecosystems. 72:101–120.

Gaskell, M., M. Bolda, J. Muramoto, and O. Daugovish. 2009. Strawberry nitrogen fertilization from organic nutrient sources. Acta Horticulturae (ISHS) 842:385–388.

Gaskell, M., and R. F. Smith. 2007. Nitrogen sources for organic vegetable production. HortTechnology 17(4):431–441.

Hadas, A., and L. Kautsky. 1994. Feather meal, a semi-slow-release N fertilizer for organic farming. Fertilizer Research 38:165–170.

Hartz, T. K., and P. R. Johnstone. 2004. N Mineralization from organic fertilizers. UC Cooperative Extension, Monterey County Crop Notes, Nov.–Dec.:8–9.

———. 2006. N availability from high-N-containing organic fertilizers. HortTechnology 16:39–42.

Hartz, T. K., J. P. Mitchell, and C. Giannini. 2000. N and carbon mineralization dynamics of manures and composts. HortScience 35:209–212.

Magdoff, F., and H. van Es. 2000. Building soils for better crops, Second edition. Beltsville, Maryland: Sustainable Agriculture Network (SAN).

May, G., and M. P. Pritts. 1990. Strawberry nutrition. Advances in Strawberry Production 9:10–24.

Maynard, D. N., and G. J. Hochmuth. 1997. Plant analysis guide for sampling time, plant part, and nutrient levels of vegetable crops. Pp. 160–167 in D. N. Maynard and G. J. Hochmuth (eds.), Knott's handbook for vegetable growers, 4th edition. New York: John Wiley & Sons.

Mikkelsen, R., and T. K. Hartz. 2008. Nitrogen sources for organic crop production. Better Crops. 92(4) 16–19. Accessed at http://www.ipni.net/ppiweb/bcrops.nsf/$webindex/90DDC9214EC7DB0A8525750600529B78/$file/BC08-4p16.pdf

Mitchell, J. P., T. K. Hartz, S. Pettygrove, D. Munk, D. May, F. Menezes, J. Diener, and T. O'Neill. 1999. Organic matter recycling varies with crops grown. California Agriculture Journal 1999 (July–Aug.):37–40.

Muramoto, J., S. R. Gliessman, D. Schmida, R. Stephens, C. Shennan, and S. L. Swezey. 2004. Nitrogen dynamics in an organic strawberry production system. California Organic Production and Farming in the New Millennium: A Research Symposium, The International House, University of California, Berkeley, California. 7/15/2004. University of California Sustainable Agriculture Research and Education Program (UC SAREP). Accessed at www.sarep.ucdavis.edu/organic/CCBCposters.pdf.

Nashlöm, T., K. Kielland, and U. Ganeteg. 2009. Uptake of organic nitrogen by plants. New Phytologist 182:31–48.

Pritts, M. P., and D. Handley (eds.). 1998. Strawberry production guide for the Northeast, Midwest, and Eastern Canada. Ithaca, New York: Natural Resource, Agriculture, and Engineering Service (NRAES).

Reeve, J. R., J. L. Smith, L. Carpenter-Boggs, and J. P. Reganold. 2008. Soil-based cycling and differential uptake of amino acids by three species of strawberry (*Fragaria* spp.) plants. Soil Biology and Biochemistry 40:2547–2552.

Soil Quality Institute. 1998. Soil quality test kit guide. Soil Quality Institute, USDA ARS NRCS. Accessed at http://soils.usda.gov/sqi/assessment/test_kit.html.

Sullivan, D. M., J. M. Hart, and N. W. Christensen. 1999. Nitrogen uptake and utilization by Pacific Northwest crops. Corvallis: Oregon State University Extension Service PNW 513.

Ulrich, A., M. A. E. Mostafa, and W. W. Allen. 1980. Strawberry deficiency symptoms: A visual and plant analysis guide to fertilization. Oakland: Agricultural Experiment Station, University of California Division of Agriculture and Natural Resources.

UC IPM (University of California Statewide Integrated Pest Management Program). 2008. Integrated pest management for strawberries, Second edition. Oakland: University of California Division of Agriculture and Natural Resources. Publication 3351.

USDA National Organic Program. 2011. Subpart C: Organic production and handling requirements. § 205.203 Soil fertility and crop nutrient management practice standard. Accessed July 10, 2009, http://ecfr.gpoaccess.gov/cgi/t/text/text-idx?c=ecfr;sid=1c0d9aa7e69d8a0ee2f14082433f9b9d;grn=div8;rview=text;node=7%3A3.1.1.9.32.3.354.4;idno=7;rgn=div8;view=text;cc=ecfr.

PHOTO: KATHY KEATLEY GARVEY

CROP MANAGEMENT

7 Erosion and Sediment Management Strategies for Organic Strawberries

DANIEL C. MOUNTJOY

Sediment management deserves special attention from strawberry producers due to the high risk of erosion when strawberries are grown on sloping land. The well-drained sandy soils of coastal bluffs and hillsides are ideally suited for strawberry production and allow for early-season production and harvest access. When planted to strawberries, however, hillside fields are more susceptible to erosion than when they are planted to other hillside crops due to the timing of land preparation and planting and the common use of plastic mulch. Some of the highest agricultural erosion rates in the western United States have been documented on strawberry fields on California's Central Coast, with soil losses of up to 33 tons per acre per year (Soil Conservation Service 1984). The erosion of any farm soil jeopardizes the sustainability of production since organic matter and fine-textured soil particles are the most likely soil components to be transported off the farm and their loss reduces the nutrient- and water-holding capacity of the soil.

Erosion and the off-farm transport of sediment and associated nutrients may also cause damage to downstream properties and can threaten water quality. The California State Water Resources Control Board now requires that farmers control agricultural nonpoint source pollution, including sediment, in order to remain in compliance with the Federal Clean Water Act and the California Porter-Cologne Act. The Agricultural Water Quality Alliance (AWQA) technical partners have developed water quality planning tools and information to help growers keep their farms in compliance (see the AWQA website, listed under References/Resources at the end of this chapter).

Strawberry fields are susceptible to erosion for several reasons. Most commercial strawberry varieties are transplanted into freshly prepared ground in the fall, just before the onset of the winter rainy season. The square-topped shape of the raised beds makes them particularly vulnerable to erosion as raindrop impacts, surface erosion, and slumping eat away at the bare soil on the shoulders of the beds. If you apply plastic mulch in advance of the rainy season, the bed itself is protected from erosion, but the resulting runoff will be greater than it would have been without the mulch and may cause scouring of the furrow bottom. The layout of strawberry fields further concentrates runoff into tail ditches and down roadways. Rapid runoff from the cropped area can overwhelm the ditches' and roads' drainage capacity and cause severe gullying, especially in farms on sloping land.

Many of the cultural operations that are intended to produce an optimal berry also result in significant modification of the natural hydrologic process. Vegetative cover of the land surface is minimal at the time of transplanting, so rainfall that would normally have been absorbed by the soil is instead concentrated on the bare, compacted soil of the furrow bottom and directed from there toward sloping drainage ditches. In this chapter, we will discuss management goals and recommended practices for reducing soil erosion at its source in the field and capturing sediment before it can enter waterways and cause water quality problems downstream.

Erosion control and soil management require a whole-farm perspective that considers both production and resource protection goals. This approach focuses on the prevention of problems before they even occur. It also recognizes that runoff ignores property lines and that effective erosion prevention may require the cooperation of neighbors. Solutions may be as simple as planting barley seed on roadways or may require complex engineering of underground drainage pipes. Free, confidential technical guidance and financial assistance for erosion control projects are available from local Resource Conservation Districts and the USDA Natural Resources Conservation Service (formerly the Soil Conservation Service). Technical conservation staff from these agencies are available in every county in California. With their help, you can develop a conservation plan to fit your new or existing farm operation.

A comprehensive farm plan for sediment management on strawberry farms involves consideration of seven management goals (MGs):

- MG 1. Evaluate current rainfall, irrigation, and storm water runoff patterns on the farm to identify erosion problem areas and their causes.
- MG 2. Develop a farm layout plan prior to planting that minimizes the slope of furrows and prevents disturbances to unstable slopes.
- MG 3. Manage uncropped areas of the farm to reduce runoff and prevent soil erosion.
- MG 4. Coordinate your efforts to control sources of runoff, erosion, and sediment with those of neighboring landowners.
- MG 5. Protect unpaved farm access roads and other heavy-use areas from concentrated flows of water.
- MG 6. Capture or filter eroded sediment and detain runoff before they leave the farm.
- MG 7. Evaluate and maintain sediment management goals and recommended practices.

Each MG can be accomplished through a set of recommended practices (RPs, described below) that you can customize to suit your farm and surrounding environment. Many of the RPs require some hydrology or engineering calculations to prevent potential failure during large storms. Overall, the best strategy for preventing sediment loss, water pollution, and damage to your strawberry crop and farm soil resources is to develop a system of practices along with backup practices to use in case one strategy fails. Prevention and a good backup plan are cheaper than middle-of-the-night repairs and cleanup after winter rains.

RECOMMENDED PRACTICES

MG 1. Evaluate current rainfall, irrigation, and storm water runoff patterns on the farm to identify erosion problem areas and their causes.

First, it is important to identify locations on the farm where rainfall or irrigation runoff already causes soil loss or soil deposition. Once you identify these, trace each one upstream to find the cause of accelerated runoff and erosion. Pay special attention to off-farm sources, steep-sloped furrows, shallow soils, and areas with large concentrations of crop runoff directed to a single road. Also consider the timing of plastic mulch placement and whether existing erosion control practices such as plastic-lined channels are working effectively, as these may also have an impact on flow and erosion patterns. Review any other farming practices currently in place and identify those cultural practices that are effectively controlling runoff and erosion processes.

MG 2. Develop a farm layout plan prior to planting that minimizes the slope of furrows and prevents disturbances to unstable slopes.

Perform land smoothing to remove irregularities on the land surface before you list up new beds so that water will not be trapped or concentrated in low spots (NRCS #466). (This and other Natural Resources Conservation Service [NRCS] practice standards and practice information sheets are listed under References/Resources at the end of this chapter and are available online.)

Proper setting and arrangement of furrows are the most important practices in terms of minimizing erosion problems on sloping farmland (RCD; NRCS #557). Good furrow layout also optimizes irrigation uniformity and water conservation. Use survey instruments or hand levels to set furrow slopes for each planting block, with the following goals:

- Avoid low spots within a furrow run where water will accumulate.

- Where furrow slopes vary within a single field block, seek out an average slope to minimize extreme slopes in parts of the field.
- Avoid furrow slopes in excess of 3 percent at the top of a field block, since runoff from these furrows will contribute to increased surface flow along the entire length of the field road (figure 7.1). Furrows with slopes in excess of 3 percent are acceptable only if they are located at the bottom of the field.
- If the furrow slope exceeds 5 percent in part of a block, break the block into two blocks and set the furrow slope to less than 3 percent in each new block section.
- Follow the natural contour of the land wherever bedding, fumigation, and spraying equipment will allow (figure 7.2).

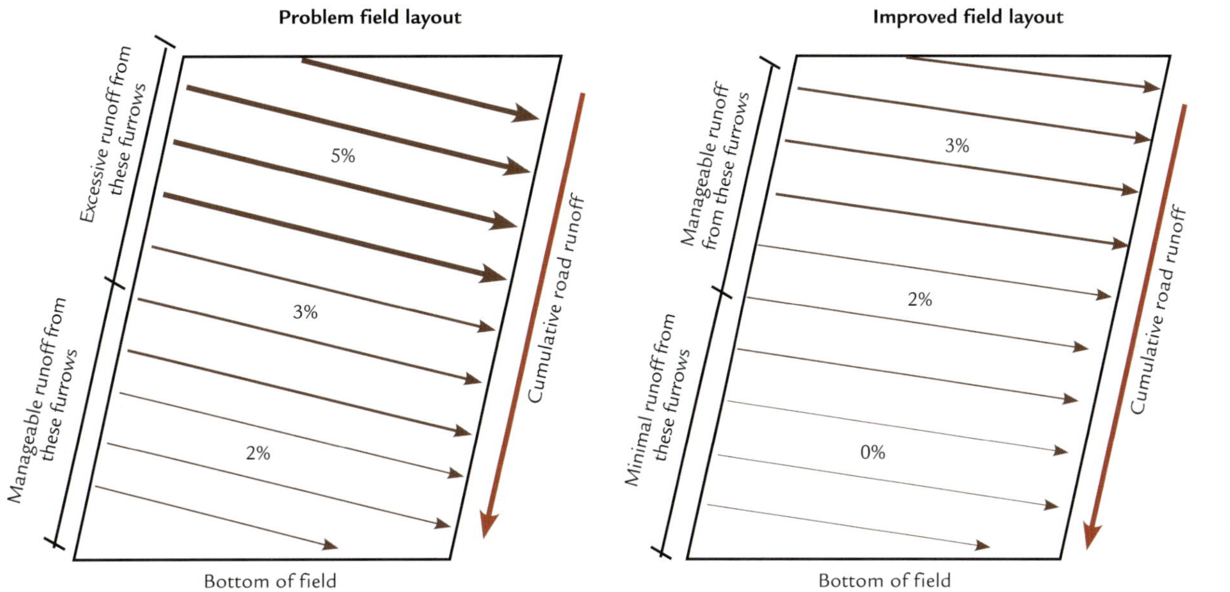

Figure 7.1. Illustration of field block layout and row arrangement on hilly lands with variable slopes. Excessive road runoff occurs in the field at left where the grade of the furrow exceeds 3% at the upper portion of field. An improved situation is achieved on the right by arranging furrows to a grade of 3% or less at the top of the field.

Figure 7.2. A strawberry row arrangement that follows the natural contour of the hillside. This arrangement results in a consistent slope to the furrow, which reduces field erosion and minimizes runoff onto the farm roads. *Photo* by L. Betts, USDA Natural Resources Conservation Service.

- Arrange the furrow layout to direct water away from the steepest farm roads.
- Avoid concentrating runoff from two field blocks onto a single roadway unless you can provide adequate drainage through an underground pipe outlet or lined channel.

The location of access roads can also make a difference. It is best to locate farm access roads between field blocks to follow the lowest water-concentrating swales and the highest ridgelines, and as needed where furrow angles must change (NRCS #560) (figure 7.3).

MG 3. Manage uncropped areas of the farm to reduce runoff and prevent soil erosion.

Cover crops can be used not only to suppress weeds and build up soil organic matter but also to protect bare soil from erosion during fallow cycles (NRCS #340). Cover crop roots, along with the cover crop residues that are incorporated later on, will increase infiltration and prevent erosion. Legume cover crops also improve soil fertility and reduce your reliance on purchased fertilizers. Winter cover crops can also be interplanted in furrow bottoms between strawberry beds to protect the soil from erosion and increase soil organic matter (figure 7.4).

Figure 7.3. A farm layout showing the placement of access roads to break up field blocks for optimal runoff management. Note that the row direction varies between most blocks to minimize the slope of furrows and that all of the blocks do not drain to the same roads. *Photo* by L. Betts, USDA Natural Resources Conservation Service.

Figure 7.4. A winter cover crop planted in furrow bottoms as an intercrop with young strawberry plants. Annual grasses such as barley can be planted in late fall after strawberry transplanting and then mowed or cultivated in spring to allow access to the crop. *Photo* by D. Mountjoy, USDA Natural Resources Conservation Service.

It is beneficial to plant bare soil and any disturbed areas of the farm, such as cut banks, field margins, and abandoned slopes, with a dense cover of low-growing vegetation to capture rainfall, control erosion, and suppress weed growth. Select a mix of species that will establish quickly and also provide long-term coverage, attract beneficial insects, and compete against weed species (NRCS #342). Another technique is to establish a hedgerow of shrubs along field margins or between field blocks (figure 7.5). This will reduce wind effects, protect slopes from erosion, and attract beneficial insects. It is best to select a mix of species that will extend the shrubs' flowering season over the strawberry growing period (NRCS #422).

You can establish larger shrubs or a row of trees to serve as a windbreak that can intercept spray drift and dispersed pest insects and weed seed. This kind of planting can also help protect steep slopes, stabilize stream or ditch banks, and reduce the incidence of soil saturation in low-lying areas (NRCS #612 and 380).

If there are gullies on the farm, it is important to prevent the gullies' advance by controlling concentrated runoff through a combination of management, vegetative, and structural measures. By installing a grade stabilization structure or pipeline at the top of the gully, you may be able to prevent further upstream head cutting and lessen the ongoing effects of erosion within the gully. Once you arrest the growth of the gully, reshape its banks and plant dense-growing grasses or shrubs to stabilize its slopes (NRCS #410, 362, 342, 620, and 557).

MG4. Coordinate your efforts to control sources of runoff, erosion, and sediment with those of neighboring landowners.

Just putting a comprehensive erosion management plan into practice on your own farm may not be enough by itself. Unless neighboring landowners follow similar practices, your efforts may not be effective. Whenever possible, work together with

Figure 7.5. A hedgerow of mixed flowering shrubs planted on the field edge is used to attract beneficial insects and protect the soil from erosion caused by rainfall impact or runoff. *Photo* by D. Mountjoy, USDA Natural Resources Conservation Service.

neighbors to solve runoff and erosion problems. Since runoff sources often do not originate on one's own land, it is crucial that you communicate and work with adjacent landowners to slow runoff and reduce the impacts of soil erosion and sedimentation downstream. Erosion problems can often be solved without great expense if neighboring landowners get involved.

MG5. Protect unpaved farm access roads and other heavy-use areas from concentrated flows of water.

First, make sure to thoroughly compact all backfill in irrigation pipeline trenches along farm roads. To minimize erosion, grade the road surface to slope slightly toward the centerline of the road rather than into ditches alongside roads, where erosion can threaten crops. Prior to winter rains, seed all erosion-prone road surfaces with annual or perennial grasses (figure 7.6). In mild winter climates, you can sow barley seed up until mid-December at rates of 180 pounds per acre, which works out to 6 pounds per 100 linear feet of 15-foot-wide road. Then mulch the seeded roads as a way to protect the soil surface from the impact of rainfall, protect the grass seeds from drying out, and prevent bird predation. For mulch, apply weed-free straw at a rate of 50 bales per acre of road surface, which works out to 1 bale per 60 linear feet of 15-foot-wide road. Crimp the straw into the soil using an open disk or a ring-shank roller (RCD; NRCS #484).

If necessary, install a diversion at the top of the field to direct runoff from adjacent areas onto roadways that have sufficient protection to handle the extra flow, or divert the runoff and dissipate its velocity of flow into an area with established perennial vegetation. This arrangement may require that you plant and maintain a strip of dense, low-growing vegetation (a filter strip) at the point where the water is diverted to dissipate the flow (NRCS #393). Another diversion technique is to construct a hillside bench, terrace, or earthen embankment midway up the field as a way to intercept runoff from roadways, reduce the length of the slope, and divert the flow across the slope to a safe outlet.

Replace existing ditches or eroding channels with a perennial grassed waterway or lined waterway to convey runoff from cropped areas, diversion terraces, or roadways. A grassed waterway is appropriate where the slope and expected channel velocities will not exceed the ability of the perennial vegetation to stabilize the soil. The waterway should be shaped and graded to accommodate the expected runoff (NRCS #412). Where vegetation and soil type are susceptible to erosion, a lined channel may be necessary. An erosion-resistant lining of rock, geotextile matting, or other permanent material will protect the channel bottom and the bases of banks from peak flows (NRCS #468).

If you expect the surface flow of runoff on farm roads to exceed 3 cubic feet per second, install one of the following conveyance systems:

- *Temporary plastic-lined channels.* Use new 4 or 6 mil plastic keyed into the soil along its sides and top edge by at least 12 inches. Install the plastic from the bottom of the slope upward, lapping each subsequent sheet by 3 feet over the top end of the lower sheet. Inspect the plastic frequently during the rainy season to check for tears or undermining (figure 7.7).
- *Underground outlet drains.* Underground outlet drains are appropriate where the road locations are permanent and concentrated surface flow is a chronic erosion problem. Contact NRCS or your local Resource Conservation District for help in determining the proper size of corrugated black plastic pipe to use, the spacing of surface riser inlets, and other construction specifications (NRCS #620).

MG 6. Capture or filter eroded sediment and detain runoff before they leave the farm.

Waterborne soil should be captured as close to the point of erosion as possible to minimize the subsequent cost of land smoothing and sediment cleanup. Plant winter grasses in the ends of furrows to filter and trap sediment that may be running off raised beds or out of the furrow bottom (figure 7.8).

Figure 7.6. Farm roads that are planted to annual grasses in the fall provide erosion protection during the winter and can be driven on in the spring. Residual grass and roots can help with dust control during the summer months. *Photo* by M. Gonzalez, USDA Natural Resources Conservation Service.

Figure 7.7. A plastic-lined channel installed down the center of a steep farm access road. Furrow runoff is discharged onto a mulched road and periodically directed into the center channel with angled waterbars. *Photo* by D. Mountjoy, USDA Natural Resources Conservation Service.

Figure 7.8. Annual grasses planted in ends of furrows slow the movement of runoff onto access roads and trap eroded sediment from the furrow bottoms that is carried in the runoff. This is especially important with full-bed plastic mulch, which accelerates rainfall runoff during winter months. *Photo* by D. Mountjoy, USDA Natural Resources Conservation Service.

At points where runoff water exits the farm and may be creating gullies on adjacent land or on the banks of natural or constructed channels, construct a long, level earthen sill to spread and evenly disperse the water. Wherever runoff from cropped areas and roadways can be dispersed evenly, plant a vegetative buffer strip or filter strip along the lower edge of the farm. For the best protection against erosion, ensure that the vegetated strips are at least 15 feet wide and are planted with a close-growing grass or legume combination to detain soils and nutrients (NRCS #587 and #393).

In those locations where you expect to see erosion and know that sediment leaves the farm, construct a sediment basin at the base of the field to intercept sediment-laden runoff (NRCS #350). A sediment basin can be designed to provide vehicle passage along the berm and can even serve as a parking or turning area after the rainy season. Consult with a hydrologist, an engineer, or an NRCS or RCD specialist to determine the scope and volume you would need for a basin capable of containing your expected sediment load and also to determine the size of the pipe outlet you would need (NRCS #638).

In some locations, sediment and excess runoff could cause gullying or flooding problems downstream. If this is the case, you will need to construct a water and sediment control basin (figure 7.9). This construction is similar to a sediment basin but is large enough to detain peak rainfall runoff as well as sediment. The detained water is then released slowly after the storm. This strategy is particularly useful on farms that use full-bed plastic mulch, which the grower can expect to increase runoff from the property. Once again, a hydrologist or specialist should be consulted to determine the optimum design (NRCS #638).

Figure 7.9. A water and sediment control basin installed at the base of a field to intercept all agricultural runoff and detain water until the sediment settles out. An outlet pipe system at the right edge of the basin is designed to release water at natural runoff rates for the land, thereby reducing the risk of flooding downstream. An emergency spillway made of rock provides overflow capacity for very large storms. *Photo* by B. Schoch, USDA Natural Resources Conservation Service.

MG 7. Evaluate and maintain sediment management goals and recommended practices.

After devising and implementing a plan, it is important that you continuously evaluate your goals and practices to ensure both the proper functioning and the effectiveness of the practices you have put in place. Correct any deficiencies as needed and remove any accumulated sediment at least once a year and spread it onto field areas after it has dried completely.

Although the topic of erosion and sediment management is presented here as a separate chapter, we need to remember that the best erosion control techniques are those that also enhance other aspects of strawberry production. Obvious examples that have been mentioned here include cover cropping for nutrient management and erosion control, field layouts that minimize furrow slope and optimize irrigation efficiency, and hedgerows planted in weedy field borders for weed and pest management and to slow and filter runoff. Upon reviewing this chapter's seven sediment management goals in terms of your own farm's situation, you may determine that your current farming activities already reduce the risk of erosion. Because the topography and setting of each farm in the landscape is unique, each farm requires its own unique sediment control practices. We have presented a framework that you can use to evaluate your own farm's situation along with a list of recommended practices. Begin by testing a few of these practices on a limited area of your farm or by contacting conservation specialists with your local RCD or NRCS office to develop a more comprehensive conservation plan specific to your farm.

REFERENCES/RESOURCES

AWQA (Agriculture Water Quality Alliance). 2008 (December 12). For farmers and ranchers: Website with information on how to assess water quality risks from farming, use conservation practices, access technical resources, and find funding opportunities. Accessed at http://awqa.org/farmers/farmersnranchers.html.

NRCS (Natural Resources Conservation Service). National Conservation Practice Standards – NHCP. (n.d.). Practice Standards and Information Sheets, eFOTG (Electronic Field Office Technical Guide), Davis, California. Accessed at www.nrcs.usda.gov/technical/standards/nhcp.html):

- NRCS #340 Cover crop and green manure crop.
- NRCS #342 Critical area planting.
- NRCS #350 Sediment basin.
- NRCS #362 Diversion.
- NRCS #380 Windbreak/shelterbelt establishment.
- NRCS #393 Filter strip.
- NRCS #410 Grade stabilization structure.
- NRCS #412 Grassed waterway.
- NRCS #422 Hedgerow planting.
- NRCS #466 Land smoothing.
- NRCS #468 Lined waterway or outlet.
- NRCS #484 Mulching.
- NRCS #557 Row arrangement.
- NRCS #560 Access road.
- NRCS #587 Structure for water control.
- NRCS #612 Tree/shrub establishment.
- NRCS #620 Underground outlets.
- NRCS #638 Water and sediment control basin.

RCD (Resource Conservation District of Monterey County), Santa Cruz County Resource Conservation District, and NRCS. 2001 (March). Controlling erosion on hillside farms. Capitola and Salinas, California. Accessed at www.rcdmonterey.org/pdf/FINALEnglishRoadsBrochure0301.pdf.

———. (n.d.). Furrow alignment: A tool for reducing soil erosion, runoff, and improving irrigation on hillside agriculture. A how-to guide for growers. Accessed at www.rcdmonterey.org/pdf/FINALEnglishFurrow.pdf.

Soil Conservation Service. 1984. Strawberry Hills Target Area Study. Davis, California: USDA.

DISEASE AND PEST MANAGEMENT

PHOTO: KATHY KEATLEY GARVEY

DISEASE AND PEST MANAGEMENT

8 Disease Management for Organic Strawberries: Foliar and Fruit Problems

JANET C. "JENNY" BROOME AND STEVEN T. KOIKE

Strawberry plants in California are subject to foliar and fruit diseases caused by a variety of pathogens. The primary fruit diseases are gray mold *(Botrytis cinerea)*, anthracnose *(Colletotrichum acutatum)*, Rhizopus fruit rot *(Rhizopus stolonifer)*, Mucor fruit rot (*Mucor* spp.), leather rot (*Phytophthora* spp.), and powdery mildew *(Sphaerotheca macularis* f. sp. *fragariae)*. The main leaf diseases are angular leaf spot *(Xanthomonas fragariae)*, anthracnose *(Colletotrichum acutatum)*, common leaf spot *(Mycosphaerella fragariae = Ramularia tulasnei)*, Phomopsis leaf blight *(Phomopsis obscurans)*, leaf blotch *(Zythia fragariae)*, and powdery mildew *(Sphaerotheca macularis* f. sp. *fragariae)*. These diseases are well described in *Integrated Pest Management for Strawberries* (UC ANR Publication 3351). Continually updated control practices, including some organically compliant methods, are listed in the *UC IPM Pest Management Guidelines for Strawberries* (UC ANR Publication 3468) and are accessible online at www.ipm.ucdavis.edu. These references and resources complement the information in this chapter.

Although there are many pathogens that can cause losses to your crop, the greatest concerns for organic producers are gray mold on the fruit and powdery mildew on the leaves, flowers, and fruit. In this chapter, we will focus on these two pathogens and the diseases they cause as well as on organically appropriate management methods. We will also briefly describe several other pathogens that may affect organic strawberry fruit and foliage.

GRAY MOLD

Botrytis fruit rot or gray mold *(Botrytis cinerea)* is the most common, most serious disease of strawberry fruit in California. This disease can affect most parts of the plant and is the direct cause of reduced yield and quality. Both pre- and postharvest fruit are impacted, as is nursery production. Postharvest losses during marketing and shipping and after final sale can be severe.

Gray Mold Symptoms and Disease Cycle

The causal fungus, *Botrytis cinerea,* infects over 200 different plant species, including many important crop plants and weeds (Coley-Smith, Verhoeff, and Jarvis 1980). The pathogen can infect healthy plant tissue—primarily the flowers—but also infects newly emergent leaves and stems where it lies quiescent, waiting until the tissue dies and then sporulating and so contributing to further buildup of inoculum (Sosa-Alvarez, Madden, and Ellis 1995). The fungus can also infect damaged and, to a lesser extent, senescent and dead plant tissues. The fungus grows into the berry receptacle from various flower parts and gives rise to a characteristic calyx-end rot of the fruit. In order to infect healthy plant tissue, the fungus needs free moisture and a nutrient source, which can be pollen, free sugars on the surface of the fruit or flower, or senescent tissues (Jarvis 1964).

Infections first appear as small brown lesions around the calyx. Both green and red berries are susceptible, but the riper the fruit is and the higher it is in sugar, the more susceptible it is and the quicker it is to rot. Infected berries maintain their original shape but take on a velvety, gray-brown coat of mycelium and spores. Rotted areas are soft and mushy at first, and then become leathery and dry in the absence of high humidity.

All California strawberry varieties are susceptible to gray mold and the fungus is ever-present in the field, on transplants, and in the soil and on other plant species. Because of this, organic approaches to managing the disease include cultural practices that influence the leaf canopy microclimate, proper management of host nutrition and water status, and sanitation practices that reduce the presence of fungal inoculum.

Plant pathologists think about disease cycles and management in terms of the plant disease triangle (figure 8.1). All three points of the triangle—a suitable host, a favorable environment, and a virulent pathogen—need to be present for disease to occur. The severity of disease is affected by all three factors and their interactions.

The disease cycle in a particular field can start each season from the overwintering forms of the fungus, which include fungal structures called mycelia and sclerotia in rotted plant debris. Sclerotia are hard, black, oblong resting structures that allow *B. cinerea* to survive in soil and on crop debris. The fungus can also come into a strawberry field on the transplants or blow in as spores from plant debris in nearby fields (figure 8.2). Copious amounts of spores are produced on this tissue, and the spores then spread to new growing tissue via wind, splashing water, and human activity in the fields.

Floral infections are important in California strawberry production because we grow indeterminate varieties that bloom for many months,

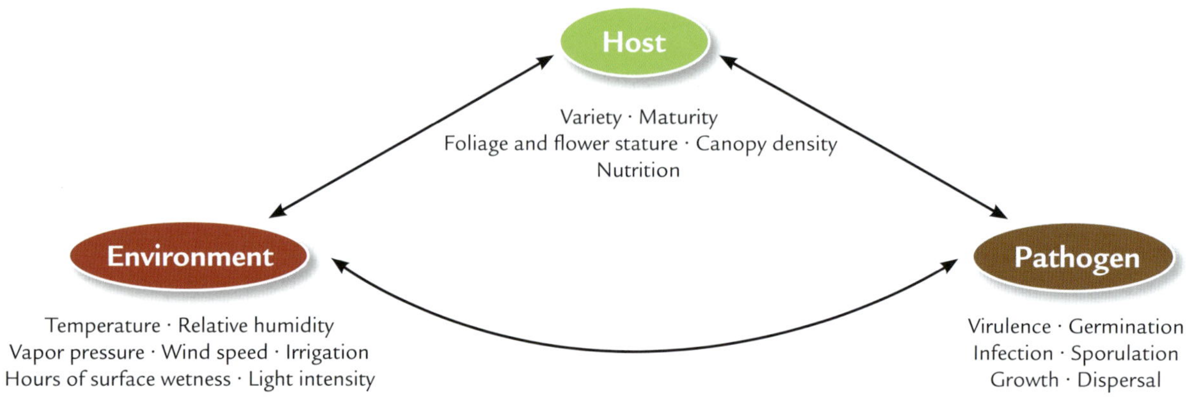

Figure 8.1. The plant disease triangle for gray mold of strawberries indicates the various factors that contribute to disease development from the host, the pathogen, and the environment.

Figure 8.2. Strawberry gray mold *(Botrytis cinerea)* disease cycle. The fungus overwinters in (A) dead plant tissue or arrives on planting material or from nearby fields; (B) spores are dispersed by wind, splashing water, and human activity directly to floral tissue or fruit; (C, D) floral tissue becomes infected throughout the season; and (E) fruit can be infected directly or a floral infection can grow into the ripening fruit. *Photos* by J. C. Broome and J. K. Clark.

providing highly susceptible tissue over long periods. The flower parts—the petals, stamens, and pistils—are the initial infection sites, and there the fungus remains dormant until the mature fruit provides enough sugars for the fungus to grow into the fleshy part of the fruit. Healthy green and red fruit can also be infected directly, but this is thought to be less common or economically important than flower infection.

Typically, the more ripe the fruit is, the more disease you can find in the field, so long as environmental conditions are suitable: wet, with moderate temperatures. However, infected fruit can remain symptom-free until a postharvest rot develops. The fungus can grow and infect plant tissue over a wide range of temperatures. Although the optimum range on strawberries is around 15° to 25°C (59° to 77°F), the fungus can grow at temperatures as low as 1°C (34°F). When temperatures are lower or higher than the optimum, the fungus requires additional hours of free moisture for infection to occur, both in grapes (figure 8.3) and in strawberries (Sutton 1998). *Botrytis cinerea* is an important postharvest pathogen as well, and can continue to grow and infect fruit at quite low temperatures. Proper postharvest handling and cooling are essential practices once the fruit is harvested (see chapter 11, *Postharvest Handling for Organic Strawberries*).

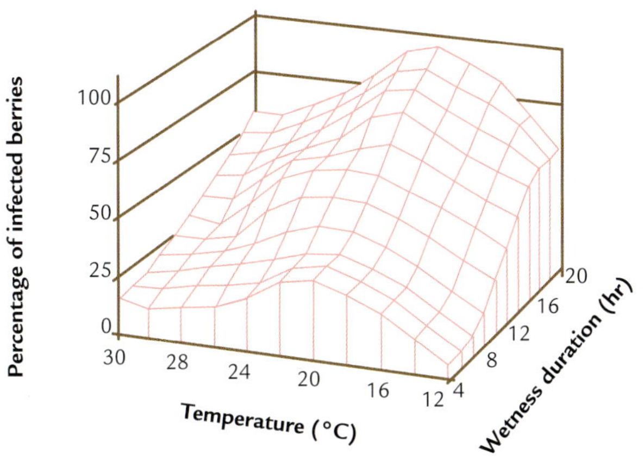

Figure 8.3. The percentage of detached grape berries infected under controlled environmental conditions at various temperature and wetness combinations (Broome et al. 1995). Temperatures between 65° and 75°F (18° and 23°C) and free moisture on the foliage from rain, dew, fog, or irrigation create ideal conditions for infection.

Research on grapes (Broome et al. 1995) demonstrated the importance of temperature and free moisture and their interaction in providing conditions suitable for *B. cinerea* infection of grape berries (figure 8.3). Similar to grapes, *Botrytis* infection of strawberry flowers at temperatures of 15° to 25°C (59° to 77°F) increased from near zero to 90 percent as the period of wetness increased from 6 to 24 hours, with 20°C (68°F) being the optimum temperature for flower infection (Bulger, Ellis, and Madden 1987). Subsequent fruit infections (up to 60% of the fruit) occurred from these flower infections, and these, too, were highest at 20°C (68°F) but with longer periods of wetness, up to 32 hours. Because of the similarity in optimum temperatures and the required period of wetness for grapes and strawberries, figure 8.3 can give you a pretty good idea of what combinations of temperature and wetness may lead to greater infection of plant tissues.

Farming System Comparison Studies

Very little research has been conducted on gray mold control for organic strawberries (Gliessman et al. 1996; Rhainds, Kovach, and English-Loeb 2002). Rhainds, Kovach, and English-Loeb found few differences in gray mold incidence rates between organic and conventionally managed strawberries, but they did find that the incidence of gray mold correlated positively with the productivity of the plants, meaning they found a higher proportion of diseased fruit associated with higher yields. This appears to support the idea that gray mold is often associated with high-yielding farming systems.

Resistant Varieties

No strawberry cultivars that are resistant to gray mold are currently available, though some do differ in their susceptibility (Legard et al. 2000). Camarosa was found to have intermediate disease levels when compared to Sweet Charlie (most disease) and Rosa Linda (least disease). In New York State, Rhainds, Kovach, and English-Loeb found the Honeoye variety to be the least susceptible to *Botrytis*, and it is considered a good early ripening cultivar for organic production there and in Europe (Rhainds, Kovach, and English-Loeb 2002; Daugaard and Lindhard 2000). Several growers have reported that some proprietary varieties appear to be less susceptible to gray mold, a factor that may be due to the more upright growth habit of their flowers and fruit. An upright stature could create environmental conditions (such as greater air movement and sunlight penetration in the plant canopy) that reduce the severity and incidence of gray mold.

Cultural Practices and Controls

Starting with clean planting material is important for control of numerous diseases, including gray mold. For organic growers, a hot water dip of the transplants may be helpful. Cultural practices such as the removal of senescent foliage and diseased fruit (Mertely et al. 2000) and the use of wider plant spacings (Legard et al. 2000) have been found to help control gray mold in Florida. The removal of senescent foliage reduced disease by 10 to 20 percent; however, researchers did not find that removal of discarded fruit in furrows significantly reduced the incidence of gray mold. It would be valuable in the future to test the impact of sanitation practices for disease control in organic production systems in California. Even with this reduction, though, the Florida researchers felt this approach was not sufficient to control gray mold without the additional use of synthetic fungicides. These findings indicate that cultural practices by themselves may not prevent gray mold development in organic plantings. Plant densities evaluated included 23, 30, 38, and 46 cm (9.1, 11.8, 15, and 18.1 inch) spacings in two offset rows per bed (plastic-mulched raised beds on 1.22 m [48-inch] centers). In general, lower-density plantings had less disease. Marketable yields are also reduced, of course, by lower plant density.

In conventional and alternative farming systems, nutrient management has been shown to influence various diseases (van Bruggen 1995), including gray mold (Daugaard 2000; Daugaard, Sorensen, and Loschenkohl 2003; Maas 1998). Excessive synthetic nitrogen use in strawberries has been found to increase disease by 60 to 80 percent when environmental variables are favorable for disease (Maas 1998). In addition to the direct effect of fertility on host susceptibility, excessive use of fertilizers can result in excess plant canopy, which then further increases the severity and incidence of Botrytis. Many organic strawberry fertilization programs do not cause excessive development of lush, dense canopies. A more open plant architecture increases airflow and sunlight penetration and decreases the buildup and duration of humidity, and appears in this way to reduce gray mold.

Excess irrigation can influence both host tissue water status and the microclimate around susceptible tissue, both of which can increase Botrytis diseases. California growers already use the appropriate irrigation method—drip irrigation, which avoids wetting flowers and fruit—and that greatly reduces the risk of gray mold fruit infections (Bulger, Ellis, and Madden 1987). The importance of wetness explains why gray mold is usually most damaging following a rain event.

Another cultural practice that reduces disease is the use of protected culture, which involves placing plastic tunnels over the strawberry beds. Researchers have found the mean incidence of Botrytis fruit rot to be 88 to 94 percent lower in tunnels than in the open field (Xiao et al. 2001). The incidence of Botrytis fruit rot for the unsprayed control in tunnels was less than 2 percent, 89 percent lower than that of the 7-day Captan schedule in uncovered beds. Their data suggest that in Florida, Botrytis fruit rot can be controlled effectively in a tunnel production system without the use of fungicides. The tunnels completely exclude rain from the plants and also impact the length and frequency of dew events, increase temperature, and reduce relative humidity. In the tunnels, there were only 30 days with an average of 4.5 hours of free moisture, as compared to 77 days with an average of 12.9 hours of wetness (free moisture) just outside the tunnels. However, while gray mold was reduced in the tunnels, the researchers observed a higher incidence of powdery mildew inside the tunnels than outside.

Researchers in Southern California have evaluated the use of tunnels for California strawberry production and confirmed the Florida results in that the tunnels increased soil and air temperature, reduced wetness by excluding rains and shortening dew periods, and reduced gray mold incidence (Shaw and Larson 2003–2004). These production systems also increased early-season and overall yield to levels that surpassed those of noncovered production systems. However, challenges remain in terms of the high cost of such systems and the establishment and maintenance of the structures in high-wind areas. These systems might be worthy of further development for the use of California organic growers who seek earlier production, higher yields, and reductions in gray mold.

Some growers and consultants believe that the use of a bug vacuum for *Lygus* control might actually increase the dispersal of gray mold spores, since they have observed increases in the disease throughout fields where the bug vacuum has been used. Additional research is needed to confirm the validity of this correlation. Possible ways to reduce any negative effects that may result from bug vacuum use might include reducing the area where it is used and only vacuuming trap crops planted to draw pest insects away from strawberry fields (see chapter 10, *Arthropod Management in Organic Strawberries*, in this manual).

Gray mold is also an important postharvest disease, so proper postharvest handling and cooling are essential practices once the fruit is harvested. Handling and packaging methods that avoid damage to the fruit are important because damaged fruit is more easily infected by gray mold spores, which are generally present on fruit surfaces and in packing areas. When possible, fruit should be cooled to near 0°C (32°F) to suppress fruit respiration and fungal growth.

Biological Control

There has been considerable research into the use of biological control agents for gray mold in strawberries, although the work has not produced any product that is commercially available in California. Two soilborne fungi, *Trichoderma* spp. and *Gliocladium* spp., have shown promise for gray mold control, but because they share some of the same environmental restrictions that the pathogen does, the degree of disease control these fungi can provide is also affected by temperature and moisture limitations (Hjeljord, Stensvand, and Tronsmo 2000). Some bacterial control agents have also received attention, and two products—Serenade and Sonata (active ingredient is the bacteria *Bacillus subtilis*)—are registered for use in California on other commodities. However, research has not found the level of control they provide to be sufficient to warrant registration for use on strawberries.

Some work with biocontrol agents has shown that these organisms induce resistance in the plant rather than having a direct effect on the pathogen. In addition, compost has been found to have a similar effect, inducing resistance to a number of pathogen and insect pests in some other cropping systems. This warrants further evaluation for effectiveness in strawberry culture (Hoitink, Stone, and Han 1997).

Chemical Controls

Few fungicides are available for gray mold control. Light mineral oils (paraffinic oil) have been found to be effective against gray mold and powdery mildew, and both are organically compliant (Gubler 1994). Experimental field applications of shrimp shell chitosan have significantly reduced postharvest gray mold on strawberries (Reddy et al. 2000). Chitosan is a low-molecular-weight glucosamine polymer derived from arthropod exoskeletons such as shrimp shells. Plant essential oils have been tested against *Botrytis cinerea* in laboratory studies, and Daferera, Ziogas, and Polissiou (2002) found that oregano, thyme, dictamnus, and marjoram oils strongly inhibit fungal growth. These materials remain unproven for gray mold control in strawberries, however, and they may produce off-flavors in the fruit.

Weather-Driven Infection Risk Model

A spray forecasting model that was developed for use on grapes has helped conventional growers improve the timing of their disease control sprays and sometimes, depending on weather conditions, reduce the number of applications needed (Broome et al. 1995). Both Hoffman and Gubler (2002) and Broome et al. (2004) have evaluated the grape model for use in timing fungicide applications in strawberries, and they found that using it with a lower action threshold than is used for grapes was useful for timing fungicide sprays and provided similar levels of disease control with fewer applications during some years. The major drawback for organic growers using this approach is the lack of any organically acceptable fungicides that have any kickback or postinfection activity. Some grape growers have used weather-based infection risk models to determine when to run their air blast sprayers, empty, through their vineyards to dry out the vines. In the future, this model or other weather-driven models that predict gray mold infection risk in strawberries could aid organic growers in managing the disease.

POWDERY MILDEW

Powdery mildew is caused by the obligate parasitic fungus *Sphaerotheca macularis* f. sp. *fragariae*. Yield losses occur due to infections of flowers and fruit as well as severe foliar infections that can damage leaves and reduce photosynthesis.

Powdery Mildew Symptoms and Disease Cycle

On leaves, white patches of mycelium develop on lower and upper leaf surfaces. The patches can enlarge until they cover much of the leaf surface. The leaf edges roll up, and purple to reddish blotches appear on the lower and upper surfaces. Leaf petioles, flower trusses, flowers, and fruit may also be affected. Powdery mildew can be particularly severe in production systems that use greenhouses or plastic tunnels.

The fungus overwinters on living, infected strawberry leaves that are present in second-year strawberry fields. The California strawberry production system, however, primarily grows annual crops, so in that case the primary inoculum comes principally from infected transplants (Nelson, Gubler, and Shaw 1996) or neighboring fields. Powdery mildew produces survival structures called chasmothecia (previously called cleistothecia), but they do not seem to play an important role in this strawberry disease (Maas 1998). Disease spreads when spores come from the transplants or from neighboring fields and are dispersed by wind. Infected flowers can be covered with mycelium and become deformed or be killed, and pollen production may be reduced by the infection, resulting in poor fruit set. Immature fruit, when infected, become hard and fail to ripen

Figure 8.4. Powdery mildew *(Sphaerotheca macularis* f. sp. *fragariae)* on leaves (A, B: colonizing bottom side of leaves; C: causing leaf edges to curl up) and on fruit (D, E: note raised seeds). Powdery mildew conidia on a conidiophore (F). *Photos* by H. Su and J. C. Broome.

in a normal fashion, whereas infected ripe fruit are firm with the seeds protruding further than they do in normal, healthy fruit, giving the fruit a "seedy" appearance (figure 8.4). Any infected plant tissue may produce spores. Environmental conditions that favor infection are dry leaf surfaces, high relative humidity, and moderate air temperatures. The disease is found mostly in the coastal growing regions and northern nurseries and has little impact in inland growing regions.

Resistant Strawberry Cultivars

The planting of genetically resistant strawberry cultivars is an important disease management strategy. There appear to be no cultivars yet available in California that are completely resistant to powdery mildew. However, researchers have found significant differences in the levels of susceptibility among various cultivars (Nelson, Gubler, and Shaw 1995, 1996). When possible, use "tolerant" cultivars, which are less susceptible to the disease (figure 8.5).

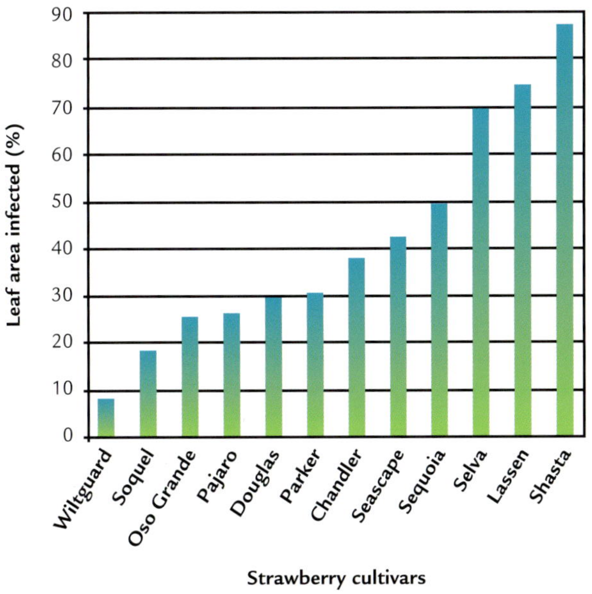

Figure 8.5. Relative susceptibility to powdery mildew for selected strawberry cultivars. Cultivar names and percentage leaf area infected: Wiltguard (8.8); Soquel (19.0); Oso Grande (26.3); Pajaro (27.1); Douglas (30.6); Parker (31.3); Chandler (38.8); Seascape (43.3); Sequoia (50.7); Selva (70.8); Lassen (76.0); and Shasta (89.0). Adapted from Nelson et al. 1996.

Cultural Practices and Controls

Proper cultural practices will not completely prevent the occurrence of foliar and fruit diseases; use of the practices described below can, however, help reduce the severity and incidence of the disease. It is important for growers to be aware that some locations have a microclimate that can favor powdery mildew development. Over time, the grower will come to know that some sites have particularly serious powdery mildew pressure; it is prudent to avoid using these locations for organic production. Bed planting density, orientation of rows, height of beds, and nutrient management (avoiding overapplication of nitrogen fertilizer) might influence powdery mildew levels; however, little research has been conducted on these factors in California.

Chemical Control

For powdery mildew, most spray materials are protectants that need to be on the plant tissue prior to infection, so timing, application method, and coverage are critical. Mineral-based products (sulfur, bicarbonates) can provide good protection in many situations if they are applied in a timely manner. Organic materials (oils, plant or compost extracts, extracts from microorganisms) have a potential to be helpful, but no field studies have demonstrated that these materials provide consistent levels of commercially acceptable control. For good organic powdery mildew control, growers currently apply sulfur seven to ten times a year and might also include three or four applications of bicarbonate materials such as Kaligreen. Light mineral oils like paraffinic oil (e.g., JMS Stylet Oil) are also organically acceptable and have shown efficacy against powdery mildew and gray mold on strawberries (Gubler 1994).

OTHER FRUIT AND FOLIAR DISEASES

Anthracnose *(Colletotrichum acutatum)*

Anthracnose, caused by *Colletotrichum acutatum*, can be a troublesome disease in strawberry production fields. It can affect leaves, fruit, petioles, crowns, runners, and roots (figure 8.6).

Figure 8.6. Symptoms of anthracnose *(Colletotrichum acutatum)* on (A, B) leaves (note salmon-colored spore masses on lesions when wet), (C) petioles, (D) immature and (E) mature fruit, and (F) salmon-colored spores forming on mature fruit. *Photos* by J. Jertberg, H. Su, and S. Koike.

Disease is favored by periods of warm, rainy weather. Fruit can be affected at any ripeness stage. Infected fruit will have small, sunken, oval-to-round brown spots if the fruit are green; if the fruit are red, the spots will be brown to black. The spots may expand to cover most of the fruit surface. Under high humidity, salmon- or orange-colored spores commonly occur on the lesions. Decayed tissue is typically firm and dry. Stem lesions appear as dark brown or black, oval, sunken spots on petioles and runners. When crown tissue is infected and becomes decayed, the entire plant may wilt and die. The internal crown tissue will be discolored cinnamon to red, distinguishing it from *Phytophthora*-rotted tissue, which is more of a chocolate brown. *Phytophthora* spp. do not cause stem or foliage lesions.

The fungus can be introduced into a field either in infested soil on planting stock or in symptom-free plant material. After transplanting, infected plants can die or become stunted, with consequent reductions in fruit yield and the direct loss of marketable fruit from berries with lesions (Mertely and Legard 2004).

The pathogen has a very wide host range that includes other crops and weeds, and as a consequence elimination of all potential inoculum may be extremely difficult (Freeman 2008; Peres et al. 2005). Gubler (2006–2007) tested a range of weeds and crops and found the fungus present in asymptomatic tissue of alfalfa, oats, bell bean, triticale, Merced rye, wheat, chickweed, lambsquarters, yellow sweet clover, hairy vetch, and nutsedge. He also found it in volunteer strawberry plants around low-elevation nurseries. In many hosts, *C. acutatum* is routinely isolated from completely asymptomatic tissue (e.g., in blueberry plants) (Yoshida et al. 2007).

Management

The keys to managing this disease are to use clean planting stock, follow good cultural procedures to keep disease inoculum from entering the field, and rotate fields to nonhost crops when possible. Right before planting, thoroughly wash all soil from the plants; this will reduce the chances of introducing infested soil into the field. In addition, dip trays of long-term cold storage (28°F) transplants into a hot water bath for 7 minutes right before planting. To prepare the plants for this treatment, thoroughly wash them to remove all dirt, then place them in a circulating water bath that is held at a constant temperature of 120°F. After that, submerge them in very cold water and then plant them as soon as possible. (This treatment is not recommended for fresh-dug transplants that have only been chilled to 33°F.)

Splashing water will spread the disease. To prevent this, only use drip irrigation in the field. Most strawberry production in California already uses drip irrigation.

Clean field equipment before you use it in your strawberry field. That will reduce the risk of transporting contaminated soil and plant parts from an infested area into a noninfested field.

Practice good weed management in and around the field to destroy any weeds that may harbor the pathogen. The pathogen can still produce spores on dead weeds, so consider physically removing them to reduce inoculum.

In warmer areas of the state, soil solarization has been shown to be effective against soilborne weeds and pathogens, including the anthracnose fungus. See chapter 9, *Cultural and Biological Management of Major Soilborne Diseases of Organic Strawberries*, for information on solarization.

Grow strawberries as an annual planting rather than a multi-year crop, since inoculum tends to build up readily in multi-year plantings. This disease tends to be much more severe in California in second-year fields.

During the growing season, look for anthracnose symptoms as part of your routine plant and pest monitoring. In particular, check after wet weather and pay close attention to developing fruit on day-neutral cultivars.

Rhizopus Fruit Rot *(Rhizopus stolonifer)*

Rhizopus fruit rot can occur on ripe fruit in the field but is seen more often on fruit after harvest. Rhizopus fruit rot first appears as discolored, water-soaked spots on the fruit. The fungus produces enzymes that leave the berry limp, brown, and leaky. Under conditions of high relative humidity, the berry rapidly becomes covered with a coat of white mycelium. The mycelium produces sporangiophores that develop black, spherical sporangia, each one containing thousands of spores (figure 8.7). When disturbed, an infected fruit can spread millions of spores through the air. Rhizopus and Mucor fruit rots look very similar to one another and may be hard to distinguish in the field.

The fungus is a saprophyte as well as a pathogen and helps break down decaying organic matter in the soil. It invades strawberries through wounds, and once there it secretes enzymes that degrade and kill tissue. The fungus survives cold periods as mycelium or spores on organic debris. The pathogen has a wide host range.

Figure 8.7. Rhizopus fruit rot. *Photo* by S. T. Koike.

Management

To reduce the incidence of this disease, use plastic mulch to reduce the amount of contact between fruit and soil. Cool the fruit as soon as possible after harvest since *Rhizopus* stops growing at temperatures below 46° to 50°F (8° to 10°C). Handle fruit with care so as not to cause damage that would create an entry point for the fungus. Make sure that when the fruit is picked, the entire fruit is removed from the stem. Do not leave behind the fleshy receptacle from the end of the fruit, as it can support fungal growth. Cultivars with thick cuticles are better able to resist Rhizopus infection. Field sanitation is extremely important: Do not leave discarded plant refuse or berries in the furrows, and remove all ripe fruit from the field at the time of harvest.

Figure 8.8. Mucor fruit rot. *Photo* by S. T. Koike.

Mucor Fruit Rot (*Mucor* spp.)

Mucor spp., like *Rhizopus* (described above), invade the strawberry fruit through wounds. The *Mucor* fungus secretes an enzyme that rapidly causes a leaky fruit rot. Under high-humidity conditions, the berry will become covered with a coat of tough, wiry mycelium and black, spherical spore-bearing structures (figure 8.8). Mucor and Rhizopus fruit rots look very much alike and may be hard to distinguish from one another in the field. However, if you use a hand lens and look at the mycelium on infected fruit, *Mucor* sporangia have a visible film of viscous liquid on the surface that makes them look sticky, whereas *Rhizopus* sporangia have a dry appearance.

Management

Use plastic mulch to keep fruit from contacting soil. Because the fungus lives on dead and decaying organic matter, field sanitation is important. Remove all ripe fruit and plant debris from the fields. Make sure that when fruit is being picked the entire fruit is removed from the stem. Do not leave behind the fleshy receptacle at the end of the fruit as it can support fungal growth. Remove and destroy all ripe and near-ripe fruit from the field after it rains. Practice good sanitation during harvest, packing, transport, and storage, and at all times avoid causing damage to fruit. Unlike *Rhizopus,* some *Mucor* species such as *M. mucedo* and *M. piriformis* are not inhibited by cold temperatures. Strawberry cultivars that have a thick cuticle are better able to resist Mucor fruit infection. Remember that rapid postharvest cooling is always important for maintaining fruit quality, even though this particular fungus is not controlled by the colder temperatures.

Leather Rot *(Phytophthora cactorum)*

All stages of strawberry fruit are susceptible to leather rot. Infected fruit develop diseased areas that range in color from brown to shades of purple. The decay often expands throughout the fruit, resulting in a brown, leathery berry—hence, the name (figure 8.9). The external infected area becomes tough, while the internal tissue is slightly softer. Infected fruit tastes bitter.

The leather rot pathogen requires splashing water to transport the zoospores (motile spores) to the fruit, or in very wet conditions (e.g., flooding, standing water, or pools of water on beds) the zoospores can swim to the plant. Watch for this disease when springtime weather is unusually wet.

Management

Leather rot is not very common on annual plantings of strawberries in California. Plantings held for 2 or 3 years, however, could be infected by the leather rot pathogen due to buildup and spread of the inoculum within the field over the course of several years.

As is true for all Phytophthora diseases, proper soil and water management are essential. Ensure that any fields chosen for production have adequate drainage to prevent standing water. Avoid overhead irrigation; use drip irrigation instead. Use plastic mulches to protect fruit from soil. Straw mulch has been effective for control of this disease in the eastern United States.

Field sanitation is important, so remove any diseased fruit from the field. If you wish to grow the planting for a second or third year and have observed leather rot during the first year, be careful to remove infected leaves and petioles so as to reduce this source of inoculum.

Soil solarization may provide control of the soilborne phase of this pathogen. See chapter 9, *Cultural and Biological Management of Major Soilborne Diseases of Organic Strawberries*, for details on management of other *Phytophthora* pathogens and details on soil solarization.

Figure 8.9. Leather rot. *Photo* by S. T. Koike.

Angular Leaf Spot *(Xanthomonas fragariae)*

Infection first appears on the lower surface of leaves as small, water-soaked spots. The lesions enlarge to form translucent, angular spots that are delineated by small veins and often exude a viscous ooze of bacteria and bacterial exudates (figure 8.10). The exudates appear as a yellowish, scaly film after drying. As the disease progresses, lesions coalesce and reddish brown spots, which later become necrotic, appear on the upper surface of the leaves. A chlorotic halo usually surrounds the infected area.

Xanthomonas fragariae is first introduced into production fields when infested transplants are planted. This bacterium overwinters in soil on previously infected plant material and is transmitted by splashing water. It is host specific and highly resistant to degradation. *Xanthomonas fragariae* bacteria can persist in debris in the soil for long periods. Lesions on the leaf surface serve as a source for secondary inoculum, which is dispersed by splashing rain, movement within the field, or overhead irrigation. The disease is favored by cool, moist days with cold nights near freezing.

Xanthomonas fragariae can cause vascular collapse of the strawberry plant, although this is uncommon in California. This symptom initially appears as a water-soaked area at the base of

Figure 8.10. Angular leaf spot *(Xanthomonas fragariae)*. (A) Typical angular lesions and bacterial ooze. With reflective light, lesions appear water soaked or dark green on the underside of leaves (B), while with transmitted light, lesions appear translucent on the upper leaf surface (C). *Photos* by J. C. Broome and C. Feliciano.

newly emerged leaves. Soon after, the whole plant suddenly dies, much like a plant infected with crown rot but without discoloration of the crown tissue. *Xanthomonas fragariae* is also associated with strawberry blossom blight in California.

Angular leaf spot generally has a minor impact on fruit yields in California. It is a concern, however, at strawberry nurseries, which may be subject to quarantine regulations for angular leaf spot on nursery stock intended for export.

Management

Use certified planting material to minimize occurrence of this disease. Crop rotations out of strawberries will reduce the pathogen's survival in the soil for future plantings. Copper-containing compounds are registered for use against the bacterium but have been known to cause phytotoxicity with repeated applications. Avoid overhead irrigation when possible.

Figure 8.11. Common leaf spot *(Mycosphaerella fragariae = Ramularia tulasnei). Photo* by S. T. Koike.

Common Leaf Spot *(Mycosphaerella fragariae = Ramularia tulasnei)*

Common leaf spot first appears as small, deep purple spots on upper leaf surfaces (figure 8.11). Spots enlarge to 1/8 to 1/4 inch (3–6 mm) in diameter with the center turning brown and then gray to white, depending on environmental conditions and the age of the leaf. A number of spots may coalesce and kill the leaf. On petioles, stolons, calyxes, and fruit trusses, elongated lesions may form and interfere with water transport in the plant, weaken the structure, or allow invasion by secondary organisms.

Common leaf spot is the most important of the fungal leaf spot diseases of strawberry in California, though economic losses from this pathogen are rare. The pathogen is introduced into fruit production fields as small, black sclerotia on infected nursery material. Germination of sclerotia is initiated by fall and winter rains or sprinkler irrigation. Spores are dispersed by wind-driven rain.

Management

Use resistant cultivars and drip irrigation, remove infected leaves when practical, and make sure your planting stock is clean. Choose a growing area with environmental conditions that are not conducive to disease development. You can use soil solarization to clean up a field that has had the disease in the past. Oso Grande is fairly susceptible to common leaf spot, as are Chandler, Pajaro, and Irvine, whereas Douglas and Parker are more tolerant.

Leaf Blotch *(Zythia fragariae)*

Strawberry leaf blotch is most often found early in the season following heavy rainfall. Symptoms consist of tan to gray lesions that quickly expand from the leaf margins on the first few leaves of the new plant. The lesions, which most often begin at leaf margins, can grow to cover one-quarter to one-half of the trifoliate leaf surface. The presence of very small black to brown fruiting bodies inside the lesions is an important indicator of this disease. This pathogen can also cause black to brown lesions on the petiole, as well as a brown blight of the calyx and decay on the calyx end of the strawberry fruit (figure 8.12). Careful and accurate diagnosis of this problem is important, as its symptoms may resemble those caused by high-salt conditions or pesticide application damage.

Zythia fragariae spreads through splashing water and is therefore much more common in winter and early spring. This pathogen survives on strawberry residue in the soil. Leaf blotch is considered a minor problem in California. The plants grow out of this disease when the winter rains stop.

Phomopsis Leaf Blight *(Phomopsis obscurans)*

This disease only occurs occasionally on California strawberries, affecting leaves and petioles. The blight first appears as purple spots on leaves. The centers then turn gray, which makes the disease hard to distinguish from common leaf spot. Older lesions have a dark brown center surrounded by a light brown inner ring and a purple or red outer ring. Spore-forming structures appear as black dots in the dark center. Lesions may grow together into large V-shaped areas with purple, red, or yellow outer edges. Symptoms that resemble those of anthracnose may appear on runners and stems. No specific control measures are warranted.

Virus and Phytoplasma Diseases

A few different virus and phytoplasma diseases can affect strawberries. For information about these, consult the second edition of *Integrated Pest Management for Strawberries* (UC IPM 2008). Viruses of concern include those that cause mottle, crinkle, mild yellow edge, vein banding, necrotic shock, and pallidosis-related decline. Two phytoplasma diseases are occasionally seen in California: lethal decline and green petal. Most of the viruses and phytoplasmas are spread by insects or through propagation of planting material. By replanting your fields every year and using certified planting material, you should be able to protect your fields from these problems. At the nursery, practices such as indexing, use of clean nursery stock produced with meristem mother plants, control of insect vectors, and removal of any symptomatic plants should keep the spread of these pathogens into production fields to a minimum.

SUMMARY

Based on an understanding of each pathogen's biology and disease cycle and the very limited available published research on organic disease control, organic disease management involves making important choices about a range of cultural practices. The first important choice is selection of a site for the strawberry planting. If possible, you should plant in a location that does not have either a history of recent cropping in strawberries or, more specifically, a history of the disease of concern, or environmental conditions that increase the likelihood of disease, such as a prevalence of heavy coastal fogs or frequent dew events. Fields that lack these qualities may have lower disease pressure.

Always choose a variety that is well suited to the region. Management practices such as sanitation are also important.

Start with clean planting stock from a well-known, certified nursery. In areas where a particular disease is known to occur, consider using a variety with lower susceptibility to that disease, such as one that resists powdery mildew. Lessons from grape and strawberry research suggest that the grower needs to grow the crop in a way that manages the microclimate, which means that cultural practices should enhance aeration and light infiltration to the strawberry flowers and fruit.

Figure 8.12. Leaf blotch *(Gnomonia comari* [anamorph *Zythia fragariae*]) on (A) leaves, (B) fruit, and (C) calyx (calyx blight). *Photos* by S. T. Koike and C. Feliciano.

When planting, consider row orientation, planting density, the stature of the variety, and bed coverage. Use plastic mulches that prevent berry-soil contact.

Remove dead or infected plant material as soon as possible. In particular, remove infected fruit after rains and remove or incorporate all plant residues after harvest. Provide appropriate amounts of nitrogen, since excess nitrogen in plant tissue has been associated with increased susceptibility to gray mold and powdery mildew.

Follow the usual practice for California strawberries and use drip rather than overhead sprinkler irrigation, and plant into soils that drain well.

Keep in mind that effective, organically compliant chemical treatments are limited. For powdery mildew, they include sulfur, light oils, and bicarbonates. Soil solarization can be an important tool for control of soilborne phases of fruit and foliar pathogens and is discussed in some detail in chapter 9, *Cultural and Biological Management of Major Soilborne Diseases of Organic Strawberries.*

REFERENCES/RESOURCES

Broome, J. C., J. T. English, J. J. Marois, B. A. Latorre, and J. C. Aviles. 1995. Development of an infection model for Botrytis bunch rot of grapes based on wetness duration and temperature. Phytopathology 85:97–102.

Broome, J. C., J. F. Strand, J. Bryer, S. L. Swezey, E. J. Morgan, and F. O. Martino. 2004. Validation of a grape *Botrytis cinerea* infection risk index for California strawberries. American Phytopathological Society Annual Meeting, July 31–August 4, 2004, Anaheim, California. Phytopathology 94:S11. Publication no. P-2004-0069-AMA.

Bulger, M. A., M. A. Ellis, and L. V. Madden. 1987. Influence of temperature and wetness duration on infection of strawberry flowers by *Botrytis cinerea* and disease incidence of fruit originating from infected flowers. Phytopathology 77:1225–1230.

Coley-Smith, J. R., K. Verhoeff, and W. R. Jarvis. 1980. The biology of Botrytis. New York: Academic Press.

Daferera, D. J., B. N. Ziogas, and M. G. Polissiou. 2002. The effectiveness of plant essential oils on the growth of *Botrytis cinerea, Fusarium* sp., and *Clavibacter michiganensis* subsp. *michiganensis.* Crop Protection 22:39–44.

Daugaard, H. 2000. Effect of cultural methods on the occurrence of grey mould (*Botrytis cinerea* Pers.) in strawberries. Biological Agriculture and Horticulture 18(1):77–83.

Daugaard, H., and H. Lindhard. 2000. Strawberry cultivars for organic production. Gartenbauwissenschaft 65:213–217.

Daugaard, H., L. Sorensen, and B. Loschenkohl. 2003. Effect of plant spacing, nitrogen fertilization, postharvest defoliation, and finger harrowing in the control of *Botrytis cinerea* Pers. in strawberry. European Journal of Horticultural Science 68:77–82.

Freeman, S. 2008. Management, survival strategies, and host range of *Colletotrichum acutatum* on strawberry. Hortscience 43:66–68.

Gliessman, S. R., M. R. Werner, S. L. Swezey, E. Caswell, J. Cochran, and F. Rosado-May. 1996. Conversion to organic strawberry management changes ecological processes. California Agriculture Journal 50:24–31.

Gubler, W. D. 1994. Use of JMS stylet oil for insect and disease control in strawberry. California Strawberry Commission Annual Report. Accessed at www.calstrawberry.com/research_library/94-0b.pdf.

———. 2006–2007. Epidemiology and control of strawberry anthracnose, and biology of *Colletotrichum acutatum* in California. California Strawberry Commission Annual Report. Accessed at www.calstrawberry.com/fileData/docs/Research_Annual_Report_2006_07_Final.pdf.

Hjeljord, L. G., A. Stensvand, and A. Tronsmo. 2000. Effect of temperature and nutrient stress on the capacity of commercial *Trichoderma* products to control *Botrytis cinerea* and *Mucor piriformis* in greenhouse strawberries. Biological Control 19:149–160.

Hoffman, L. E., and W. D. Gubler. 2002. Validation of a Botrytis infection risk model: Part I and Part II. California Strawberry Commission Pink Sheets. Accessed at www.calstrawberry.com/research_library/02-17_gubler_bot%202002_part%20one.pdf, http://www.calstrawberry.com/research_library/02-18_gubler_botrytis_part_two.pdf.

Hoitink, H. A. J., A. G. Stone, and D. Y. Han. 1997. Suppression of plant diseases by composts. HortScience 32:184–187.

Jarvis, W. R. 1964. The effect of some climatic factors on the incidence of grey mould of strawberry and raspberry fruit. Horticultural Research 3:65–71.

Legard, D. E., C. L. Xiao, J. C. Mertely, and C. K. Chandler. 2000. Effects of plant spacing and cultivar on incidence of Botrytis fruit rot in annual strawberry. Plant Disease 84:531–538.

Maas, J. L. (ed.). 1998. Compendium of strawberry diseases, Second edition. St. Paul, Minnesota: American Phytopathological Society.

Mertely, J. C., C. K. Chandler, C. L. Xiao, and D. E. Legard. 2000. Comparison of sanitation and fungicides for the management of Botrytis fruit rot of strawberry. Plant Disease 84:1197–1202.

Mertely, J. C., and D. E. Legard. 2004. Detection, isolation, and pathogenicity of *Colletotrichum* spp. from strawberry petioles. Plant Disease 88:407–412.

Nelson, M. D., W. D. Gubler, and D. V. Shaw. 1995. Inheritance of powdery mildew resistance in greenhouse-grown versus field-grown California strawberry progenies. Phytopathology 85:421–424.

———. 1996. Relative resistance of 47 strawberry cultivars to powdery mildew in California greenhouse and field environments. Plant Disease 80:326–328.

Peres, N. A., L. W. Timmer, J. E. Adaskaveg, and J. C. Correll. 2005. Lifestyles of *Colletotrichum acutatum*. Plant Disease 89:784–796.

Reddy, M. V. B., K. Belkacemi, R. Corcuff, F. Castaigne, and J. Arul. 2000. Effect of preharvest chitosan sprays on postharvest infection by *Botrytis cinerea* and quality of strawberry fruit. Postharvest Biology and Technology 20:39–51.

Rhainds, M., J. Kovach, and G. English-Loeb. 2002. Impact of strawberry cultivar and incidence of pests on yield and profitability of strawberries under conventional and organic management systems. Biological Agriculture and Horticulture 19:333–353.

Shaw, D. V., and K. D. Larson. 2003–2004. Pomology: Strawberry genetics, breeding, physiology, and production management. California Strawberry Commission Annual Report, pages 136–144. Accessed at www.calstrawberry.com/fileData/docs/CSC_Annual_Research_Report_2003-2004.pdf.

Sosa-Alvarez, M., L. V. Madden, and M. A. Ellis. 1995. Effects of temperature and wetness duration on sporulation of *Botrytis cinerea* on strawberry leaf residues. Plant Disease 79:609–615.

Sutton, J. C. 1998. Botrytis fruit rot (gray mold) and blossom blight. Pp. 38–31 in J. L. Maas (ed.), Compendium of strawberry diseases, Second edition. St. Paul, Minnesota: American Phytopathological Society Press.

UC IPM (University of California Statewide Integrated Pest Management Project). 2008. Integrated pest management for strawberries, Second edition. Oakland: University of California Division of Agriculture and Natural Resources. Publication 3351.

van Bruggen, A. H. C. 1995. Plant disease severity in high-input compared to reduced-input and organic farming systems. Plant Disease 79:976–984.

Xiao, C. L., C. K. Chandler, J. F. Price, J. R. Duval, J. C. Mertely, and D. E. Legard. 2001. Comparison of epidemics of Botrytis fruit rot and powdery mildew of strawberry in large plastic tunnel and field production systems. Plant Disease 85:901–909.

Yoshida, S., T. Tsukiboshi, H. Shinohara, M. Koitabashi, and S. Tsushima. 2007. Occurrence and development of *Colletotrichum acutatum* on symptomless blueberry bushes. Plant Pathology 56:871–877.

PHOTO: PENNY LEFF

DISEASE AND PEST MANAGEMENT

9 Cultural and Biological Management of Major Soilborne Diseases of Organic Strawberries

KRISHNA V. SUBBARAO, FRANK N. MARTIN, AND STEVEN T. KOIKE

PATHOGENS AND DISEASE SYMPTOMS

A number of soilborne diseases pose significant challenges to organic strawberry production in California. Two of these, Verticillium wilt and Phytophthora crown and root rot, are the most widely found and most destructive, and hence the most important soilborne diseases for strawberry growers. A number of other soilborne pathogens can also be problematic for organic strawberry production, but for the most part they are less common or cause only limited crop losses. These other pathogens include species of *Colletotrichum*, *Pythium*, *Rhizoctonia*, *Cylindrocarpon*, *Macrophomina*, and *Fusarium*. Symptoms and damage caused by *Verticillium*, *Phytophthora*, and the other soilborne pathogens may all be very similar in appearance. The diseases therefore must be carefully diagnosed. Precise identification of the soilborne pathogen will usually require laboratory analysis and tests.

Two Major Concerns

Both Verticillium wilt and Phytophthora root and crown rot cause poor growth and even death of plants, and no effective resistance is yet available in commercial strawberry cultivars. Verticillium wilt (figure 9.1) is caused by the fungus *Verticillium dahliae*, which is widely distributed in the agricultural soils of California and affects such diverse crops as artichoke, cotton, lettuce, pepper, pistachio, potato, strawberry, tomato, and watermelon, as well as a number of crucifer crops. The pathogen persists in the soil for very long periods by producing tiny resting structures called microsclerotia (figure 9.1). In response to nutrients released by the root of a potential strawberry host, microsclerotia will germinate, grow toward the root, and infect the plant.

Figure 9.1. Verticillium wilt symptoms in strawberry (top) and microsclerotia magnified 1,000 × (bottom). *Photos* by S.T. Koike and K. Subbarao.

Strawberries are especially susceptible to *V. dahliae*: as few as two microsclerotia per gram of soil can result in 100 percent incidence of the disease (Harris and Yang 1996). The long strawberry growing season also leaves the plant exposed to soilborne inoculum for an extended period. Infected plants show varying degrees of poor growth, stunting, dieback of older leaves, lack of fruiting, plant collapse, and finally plant death. After an infected plant dies, the pathogen can continue to produce microsclerotia inside the dead plant, and these will return to the soil when crop residues are incorporated into the soil after harvest.

Note that strains of the Verticillium wilt pathogen that originate from other crops may or may not affect strawberries. Researchers have determined, for example, that *V. dahliae* from lettuce can infect strawberries and *V. dahliae* from strawberries can infect lettuce, but *V. longisporum* from cauliflower apparently will not infect strawberries or lettuce. Similarly, *V. dahliae* from peppers will not infect strawberries. These differences in host range underscore the importance of knowing your field's history of Verticillium wilt and of crops that may or may not have developed the disease in previous seasons.

The development of Verticillium wilt can be strongly influenced by timing of infection, cultivar susceptibility, and environmental conditions. Plants may be severely affected if infection occurs very early in the season, the infected cultivar is highly susceptible, or infected plants are subject to periods of environmental stress from high temperatures or a shortage of water. Disease symptoms may be slower to develop in less-susceptible cultivars or when infection occurs later in the season. Late infections are typical when inoculum levels are low because the root system has had a chance to develop extensively before encountering the pathogen in soil. In such cases, infected plants may be indistinguishable from healthy plants for much of the season and yield reductions may be minimal.

Several *Phytophthora* species cause crown and root rot diseases in strawberry: *Phytophthora cactorum*, *P. citricola*, and *P. fragariae*. Phytophthora crown and root rot caused by *P. cactorum* is probably the most common Phytophthora disease in California strawberries (Winterbottom et al. 1996). Plants typically are stunted, have small leaves at disease onset, produce few fruit, and can rapidly collapse (figure 9.2). A key symptom is the internal brown discoloration of roots and crowns. If rains occur during fruiting, *P. cactorum* can also cause a fruit rot of strawberry. Red stele disease caused by *P. fragariae* and other Phytophthora diseases is much less prevalent on strawberries in California.

Figure 9.2. Phytophthora crown rot symptoms showing plant collapse (top) and discoloration of internal crown tissue (bottom). *Photos* by S. T. Koike.

Phytophthora species are soilborne pathogens that also produce resting structures (oospores) that enable them to survive for long periods in soil. The development and severity of Phytophthora diseases is greatly influenced by the amount of water in the soil. When the soil is saturated, the pathogen produces and releases many zoospores, microscopic swimming spores that move through water-filled pores and infect strawberry roots. Infections are favored by cool to moderate temperatures, which are typical in coastal strawberry growing areas.

Additional Soilborne Challenges

Besides *V. dahliae* and *Phytophthora* spp., several other pathogens cause crown or root rot diseases of strawberry. Anthracnose can be caused by any of a number of species of the fungus *Colletotrichum*, though the most important species in California is *C. acutatum* (Gunnell and Gubler 1992). Anthracnose may be more familiar as an aboveground disease that results in stem, leaf, and fruit lesions, but it can also damage belowground parts of plants. Anthracnose symptoms usually develop in production fields where the transplants were either contaminated with or actually infected by the pathogen before transplanting. Affected roots are rotted, and the internal crown tissue is discolored and shows sections that are brown. Plants with root and crown infections are slow to grow and establish, remain stunted, and have foliage that wilts and dies. Severely affected plants will die. Spores of the anthracnose fungus, formed on the crown and lower parts of plants, can be splashed by rain up onto foliage and fruit, causing lesions to form there. While *C. acutatum* can survive in soil, the exact biology of this pathogen under California production field conditions has not yet been determined.

While not lethal, another of the more common root diseases encountered in nonfumigated soil is caused by a complex of pathogens and is known as black root rot. The disease can result in necrotic lesions on the root, reduced root systems, smaller plants, reduced shoot growth, and reduced yields, but generally not plant mortality. High soil moisture tends to enhance the severity of the disease.

In other strawberry production areas in the world, black root rot has been attributed to various combinations of root infection by *Pythium* spp., three different anastomosis groups (AGs) of binucleate *Rhizoctonia* spp., *Cylindrocarpon* spp., and the lesion nematode *Pratylenchus penetrans* (reviewed in Wing, Pritts, and Wilcox 1994). In Connecticut, researchers believed that binucleate *Rhizoctonia* spp. and the lesion nematode are the primary causes of the disease complex (LaMondia and Martin 1989).

In the strawberry production areas of California's Central Coast that have been examined for the disease, researchers have not identified the lesion nematode as a problem. However, all of the indicated fungal pathogens have been recovered from necrotic roots of plants growing in poorly fumigated or nonfumigated soils, including the same three AGs of binucleate *Rhizoctonia* spp. that have been associated with the disease in other strawberry production areas (Wilhelm et al. 1972; Yuen et al. 1991; Martin 2000). While multinucleate *Rhizoctonia solani* has been recovered from infected plants, the binucleate *Rhizoctonia* spp. are more commonly encountered (Martin 2000, 2001). Multiple *Pythium* spp. have been recovered, but the broad host range species *P. ultimum* and *P. irregulare* are the most commonly encountered.

Very little research has been done on root infection by *Cylindrocarpon* spp. While commonly recovered from symptomatic tissue, pathogenicity tests with *Cylindrocarpon* isolates have only resulted in limited damage to roots (Martin, unpublished). More research will be needed to fully characterize these pathogens that are associated with black root rot disease complex, understand their specific impacts on yield and quality, and explore management options for organic strawberry production.

Charcoal rot or Macrophomina crown rot is caused by the soilborne fungus *Macrophomina phaseolina*. This is a relatively new problem for production strawberries in California (Koike 2008). Symptoms consist of plant stunting, wilting of foliage, drying and death of older leaves, reduced fruit production, and the eventual collapse and death of plants (figure 9.3). When you examine the internal crown tissues, the vascular and cortex tissues appear dark to orange-brown (figure 9.3). Diseased roots are completely black and are soft and rotted. No fruiting bodies or other fungal structures have been observed on diseased plants. The pathogen produces numerous dark, irregularly shaped sclerotia that enable it to survive in the soil. This disease usually occurs late in the production season when plants are subject to stress factors such as warmer, elevated temperatures and heavy fruit loads.

Fusarium wilt of strawberry is another recently reported problem in California (Koike, Kirkpatrick, and Gordon 2009). Symptoms are very similar to those caused by charcoal rot and consist of wilting of foliage, drying and withering of older leaves, stunting of plants, and reduced fruit production (figure 9.4). Plants can eventually collapse and die. Internal vascular and cortical tissues of plant crowns also show a brown to orange-brown discoloration (figure 9.4). Fusarium wilt is caused by the soilborne fungus *Fusarium oxysporum*, sometimes listed as *F. oxysporum* f. sp. *fragariae*. Apparently, this pathogen

Figure 9.3. Charcoal rot symptoms showing plant collapse (top), discoloration of internal crown tissue (middle), and rotted, reduced root system (bottom; healthy plant on left). *Photos* by S. T. Koike.

Figure 9.4. Fusarium wilt symptoms showing plant collapse (top) and discoloration of internal crown tissue (bottom). *Photos* by S. T. Koike.

is only able to infect strawberries and produces structures (chlamydospores) that enable it to survive in soil for long periods. Additional information on soilborne diseases of strawberry is available in *Integrated Pest Management for Strawberries, Second Edition* (UC ANR Publication 3351), and the *Compendium of Strawberry Diseases, Second Edition* (American Phytopathological Society Press).

Nematodes are microscopic, aquatic roundworms that cause problems for many crops. Most nematode damage comes when large populations develop on the root system of a crop and cause damage to small feeder roots, interfering with the function of the root system, and so cause stress to the plant. A few nematode species that can be found aboveground infect leaves. For organically produced strawberries, however, harmful economic impact from nematodes is rarely seen and has not been documented.

Regardless of this, though, growers should remain aware that nematodes such as the leaf-infecting foliar nematode *Aphelenchoides fragariae* and the root-infecting northern root knot nematode *Meloidogyne hapla* can infest strawberries. As indicated above, researchers have observed situations in which a complex of fungal pathogens and root-infecting nematodes could result in stress, negative symptoms, and problems for various crops, including strawberries. Symptomatic plants and those developing problems should therefore be carefully examined and diagnosed for possible fungal, bacterial, viral, and nematode causal agents. Refer to the *UC IPM Pest Management Guidelines for Strawberry* (UC ANR Publication 3468) for additional information on nematodes.

CULTURAL CONTROLS

Cultural controls form a major component of a sustainable agricultural system (Conway 1996) and are important tools for minimizing disease losses in many crops (Menzies and Belanger 1996). Cultural controls involve following these principles:

- Use plants that are resistant or less susceptible to the pathogen.
- Take steps to avoid or reduce pathogen populations.
- Create growing conditions that favor growth of the plant while discouraging pathogen development.

Resistant Strawberry Cultivars

True genetic resistance to *Verticillium* and *Phytophthora* pathogens is not yet available for commercial strawberry cultivars, but recently released cultivars are for the most part less susceptible than older cultivars. For example, experimental inoculations have shown that the Camino Real and Albion cultivars sustain less damage from Verticillium wilt than Camarosa or Diamante. Preliminary research indicated that the Seascape and Chandler cultivars may be tolerant to *Macrophomina* and that Ventana may be less susceptible to *Fusarium* (Koike and Gordon, unpublished). Cultivars also show differential resistance to other pathogens such as *Phytophthora, Pythium,* and *Rhizoctonia* (see chapter 3, *Cultivar Selection for Pest Management in Organic Strawberry Production,* for more on host resistance and disease management). It is likely that the various strawberry breeding programs being conducted in California will continue to develop cultivars with increasing levels of resistance.

Site Selection

The choice of an appropriate planting site for organic strawberries is extremely important. Without sufficient planning, you may end up placing the crop in a field that has a history of soilborne disease, a previous strawberry crop (which might increase the risk of disease, including some foliar diseases), or other previous crops that are susceptible to and harbor pathogens that harm strawberries (Jordan, Sneh, and Eddy 1972; Milholland 1994; Pettitt and Pegg 1994). Less than optimal locations also include fields that are adjacent to an external source of pathogens or are in places where the environment does not favor growth and development of the strawberry crop.

Select the site based on a careful examination of the field's history. Consult records and interview personnel on all pertinent production aspects of the field: overall cropping history, history of strawberry plantings, previous production practices, and soil inputs (e.g., amendments, composts, and cover crops). Find out the exact history of previous diseases in the field, especially soilborne problems, for all crops. Growers are advised to obtain laboratory results that could document soil pathogen populations prior to planting, provided that this information for the field is available from previous tests. This type of laboratory testing is expensive and is not required each season or year unless there was a major disease problem the previous year, but it is a good field management practice to obtain these results at regular intervals.

Determine whether the characteristics of the soil (e.g., type, condition, variability, depth, fertility, and salinity) and the water source (e.g., availability, quality, salinity, drainage, runoff, and destination) are amenable to strawberry production. With regard to environmental conditions, find out if the weather and microclimate patterns at the site have any peculiarities that would adversely affect strawberry production.

Finally, collect information about the area surrounding the prospective planting site. Ask the same questions as you asked for the site itself, but now with regard to adjacent and nearby fields: What are the historical and current circumstances of adjacent fields? Of the surrounding area? Have strawberries been heavily cropped in the area? Has Verticillium wilt of strawberry or lettuce been confirmed nearby or has Verticillium wilt only been observed on crops for which the pathogen has a narrow host range restricted to that crop (e.g., cauliflower or pepper)?

Pathogen Exclusion

In principle, the best way to control a disease is to prevent the pathogen from entering the field in the first place. Following this logic, the best practice is to use transplants that are healthy and as free as possible from pathogens and disease, though in the case of strawberries the complete exclusion of pathogens is not possible.

Because soil and water can harbor soilborne pathogens, avoid introducing infested soil or water into your field. Soil and mud adhering to tractor equipment and implements can readily spread soilborne pathogens into previously clean fields and areas. Insofar as possible, do not allow runoff water from external sources to enter the field.

For charcoal rot and Fusarium wilt diseases, in-field movement of infested soil accounts for spread within fields. Initial occurrence of the diseases is often restricted to distinct sections or corners of a field. As soil is moved by tractors, the pathogen spreads to other areas of the same field; collapse of subsequent strawberry plantings in larger patches throughout the field demonstrates how the pathogen has been spread.

Proper Soil, Nutrient, and Water Management

Prepare the field so as to enhance the growth of strawberries and discourage the activity of soilborne pathogens. Design and form beds in a way that enhances drainage and soil structure. Use appropriate fertilizers and irrigation methods. Avoid planting in fields where extensive crop residue is still present.

It is important to carefully manage soils in an organic system; there is evidence that some factors in such systems may inhibit pathogens and reduce soilborne diseases. Van Bruggen (1995) recognized trends when she compared soil biological and chemical characteristics and soilborne diseases in organic versus conventional farming systems. Soils from organic and reduced input systems had a thicker topsoil layer, lower bulk density, and greater water-holding capacity than soils from conventional farms. Soil nitrate content tended to be lower and the nitrogen mineralization rate higher in organic than in conventional systems. Organic matter content and microbial biomass and activity were also generally higher. Increased use of organic amendments may reduce root diseases by increasing the level of soil microbial activity and subsequent competition and antagonism in the root zone. In another study, organically managed plots showed higher biological activity than did conventionally managed plots after 21 years (Maeder et al. 2002). The researchers presumed that this was caused by more efficient resource usage and enhanced soil floral and faunal diversity in the organic system.

Crop Rotation and Cover Crops

The benefits of rotating diverse crops to maintain crop yields and suppress soilborne diseases have been recognized and exploited for centuries (Cook 1988; Umaerus 1992). Organic production systems have relied on crop rotations to improve crop growth, manage diseases, increase profitability, and realize stability and sustainability of the farming systems. Crop rotations generally reduce pathogen populations because the pathogens that infect one crop are not always able to infect and multiply on an unrelated crop (Cook 1988, 1991). For example, none of the *Rhizoctonia* species that infect strawberry will attack broccoli, and the *Phytophthora capsici* pathogen of pepper will not attack strawberry. In one analysis, researchers suggested that the lack of strawberry root diseases at a particular farming site was attributable to its long history of Brussels sprouts production (Gliessman et al. 1996).

We do not know exactly how crop rotations affect soilborne pathogens. Crop rotations may suppress soilborne diseases by bringing about changes in microbial interactions (Lockwood 1988) or soil physical factors (Blevins, Frye, and Smith 1985; Evans and Gleeson 1973; Nederman 1991; Roberson, Sarig, and Firestone 1991; Williams 1996). Crop rotation may reduce the amount of inoculum in the soil, increase the population density of microbes antagonistic to the pathogen, or increase overall microbial competition for nutrients present in the soil or on root surfaces.

Typically, growers rotate organic strawberry fields out of strawberry production for 1 to 3 years before replanting to the same crop. The choice of crops for rotation depends on several considerations. First, you want to select crops that fit into the current cropping and marketing systems for your region. Second, you should choose crops that are not susceptible to the same pathogens that affect strawberry. Finally, if feasible, you should use rotation crops that have demonstrated an ability to suppress soilborne pathogens.

Broccoli is a commonly grown crop that has shown suppressive effects against soilborne pathogens. Broccoli is not susceptible to Verticillium wilt, and broccoli residue can reduce *V. dahliae* propagules in soil. Field experiments have consistently demonstrated the effectiveness of broccoli rotation and residue amendment for Verticillium wilt management in cauliflower (Subbarao and Hubbard 1996; Xiao et al. 1998). To achieve maximum effect, the broccoli residue should be incorporated when the soil temperature is at least 20°C (68°F). Fresh broccoli residues appear to be more effective than dried broccoli residues (Subbarao and Hubbard 1996).

The mechanisms by which the crop residues of broccoli and other brassicas act on soilborne pathogens may be chemical or biological or a combination of both. During the decomposition of crucifer residues, the breakdown of glucosinolates (a group of chemicals found in crucifer plants) releases chemical by-products that can kill soilborne pathogens (Brown and Morra 1997; Chan and Close 1987; Chew 1988; Fenwick, Heaney, and Mullin 1983; Lewis and Papavizas 1970, 1971, 1974). On the other hand, broccoli residues may reduce the presence of *V. dahliae* microsclerotia in the soil by enhancing antagonistic soil microorganisms. An analysis of soil samples from field plots with and without broccoli indicated a 1,000-fold increase in the number of actinomycetes, a 100-fold increase in total bacterial counts, and increased numbers of *Enterobacter* bacteria in plots with broccoli (Subbarao et al., unpublished). Sheer numbers may allow these microorganisms to outcompete the pathogen populations for nutrients or simply reduce the activity and growth of pathogens in the root zone (Cook 1984; Curl and Truelove 1985; Hornby 1990; Labeda 1986; Lynch 1982; Rovira, Elliot, and Cook 1990).

Recent research has tested the efficacy of strawberry rotations with broccoli and other crops for Verticillium wilt management (Njoroge et al. 2009; Subbarao et al. 2007). Broccoli rotations significantly reduced pathogen inoculum (figure 9.5) and wilt disease in strawberries (figure 9.6). After two crops of broccoli, inoculum levels were reduced by 45 percent in one organic field trial. This reduction was significantly greater than that seen for the Brussels sprouts rotation crop. Lettuce, when rotated with strawberry, may actually have increased *V. dahliae* microsclerotia in the soil (figure 9.5). Increased strawberry yields were also obtained from the broccoli rotation plots (figure 9.7).

Other crops that have been studied for their effects on suppressing plant pathogenic fungi include alfalfa, cabbage, corn, mustard, oats, pea, peanut, rapeseed, sorghum-Sudangrass, and Sudangrass. In some studies, crop residues from Sudangrass or corn were effective in suppressing Verticillium wilt of potato (Blevins, Frye, and Smith 1985), but these have not yet been shown to be effective for strawberries in California.

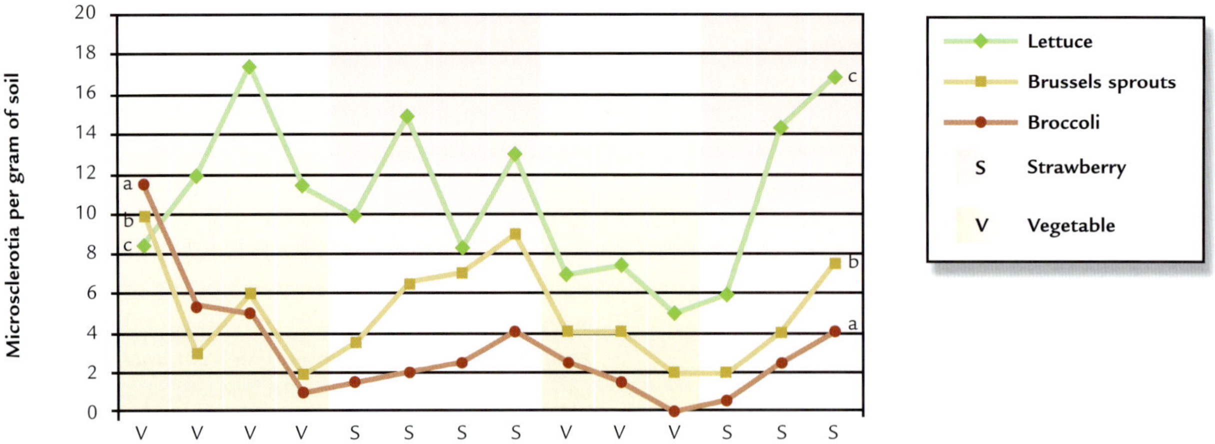

Figure 9.5. Effect of crop rotation on inoculum (microsclerotia) densities of *Verticillium dahliae* from 1997 to 2000. Soil samples taken during the course of the study in different rotation treatments. Means with the same letter are not significantly different at < 0.05, $n = 30$.

Brassica plants have been employed as cover crops to help manage diseases. Recent work with yellow and oriental mustards (*Sinapis alba* and *Brassica juncea*) has shown beneficial effects in rotation with strawberries (Lazzeri and Manici 2001). More than 15 soilborne diseases, including *V. dahliae,* have been documented in the literature as being suppressed by residue from mustard cover crops (Olivier et al. 1999). However, this disease management strategy requires further study in California before it can be employed in organic strawberry production.

Microbial Inoculants

Martin and Bull (2002) reviewed biological approaches to soilborne disease control in strawberries and described recent work with microbial inoculants in California strawberry production. While there has been only limited work, some of the results look encouraging. For example, some bacterial colonizers of the strawberry root zone can significantly enhance root and shoot growth in growth chamber trials (Martin and Bull 2002). Some of these strains also enhanced yields in a replicated field trial when they were applied as a root dip at planting, but researchers did not get the same results when the trial was repeated the following year. When the inoculants were applied through drip lines at multiple intervals during the season during a third trial, researchers observed yield increases of 20 percent (Martin 2001).

Some researchers theorize that mycorrhizae may confer bioprotection to various plants, including strawberries, against several pathogens, including *Verticillium* and *Phytophthora* species (Harrier and Watson 2004; Vigo, Norman, and Hooker 2000). Bull, Shetty, and Subbarao (2002) found that soil myxobacteria inhibited several soilborne pathogens in vitro but also found that they could interfere with some fungal biological control agents. Clearly, more work is needed on strain selection and optimization of application technologies before biological control agents can become part of a commercially viable soilborne disease management strategy for organic strawberries.

Organic Amendments and Inputs

Organic amendments such as composts contribute to the biological control of root diseases in containerized production systems (Hoitink, Stone, and Han 1997). However, you have to manage many other factors properly if you want to obtain similar results in a field soil. The raw materials that make up the compost, the composting process, the microorganisms that colonize the compost after peak heating, the chemical and physical characteristics of the compost, and the rate of compost application all affect the potential for biological control of pathogens. While amendments and composts should remain standard inputs for organic systems, these inputs have not yet demonstrated direct and unambiguous control of soilborne pathogens in commercial organic settings.

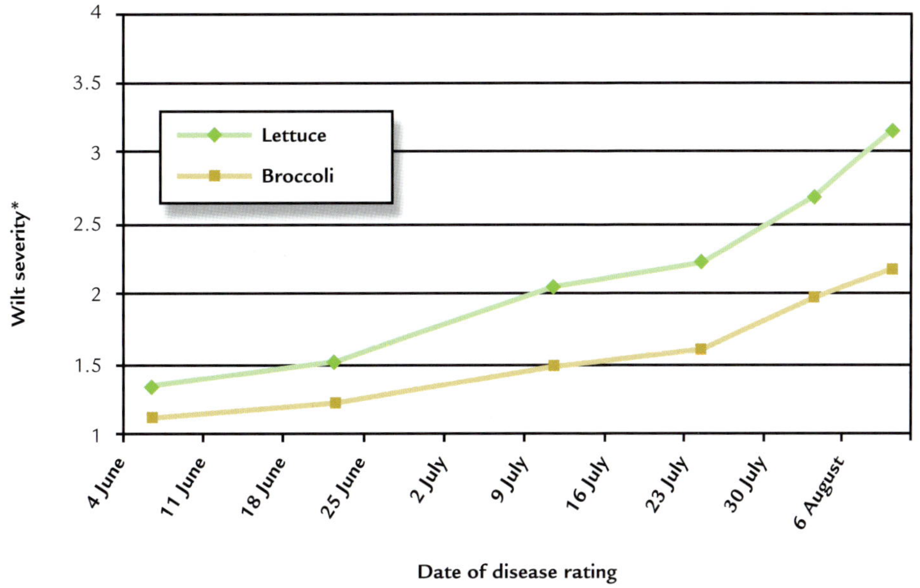

Figure 9.6. Severity of Verticillium wilt on strawberries grown in plots previously cropped with either lettuce or broccoli in an organically managed system.

* 1 = healthy plant; 2 = plant moderately stunted; 3 = plant moderately stunted and outer rosette of leaves showing some dieback; 4 = plant moderately stunted and outer rosette of leaves showing moderate dieback.

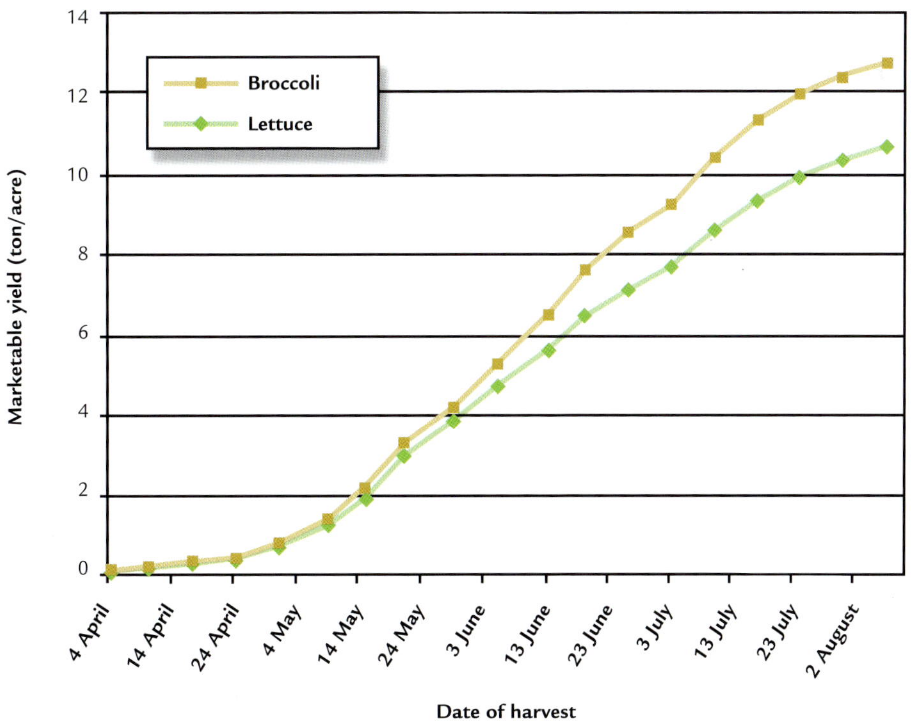

Figure 9.7. Marketable strawberry yields in the different rotation treatments under an organic production system.

CHEMICAL AND PHYSICAL CONTROL

There are few chemical options available for soilborne disease control in organic strawberries. Solarization, however, where clear plastic is laid on the soil surface to heat the soil and trap solar radiation, can be used as a preplant soil treatment to manage soilborne pathogens and pests in organic systems (Katan and DeVay 1991). It has been used with some success in field experiments for strawberries and lettuce in California (Gamliel and Stapleton 1993). Solarization improves plant development, growth, quality, and yield for several years by causing complex changes to the biological, physical, and chemical properties of the soil (Katan and DeVay 1991).

The effectiveness of this pest control tool is directly linked to climate, and in particular the amount of sunlight and the ambient temperatures during the solarization process (Stapleton 1994). Researchers have found solarization to be less effective in the Central Coast area of California, where most strawberry production occurs. By combining solarization with organic amendments such as chicken manure and cabbage residues, researchers have been able to increase control of pathogens in lettuce fields (Gamliel and Stapleton 1993) and, partially, in cauliflower (Subbarao, Hubbard, and Koike 1999). With further evaluation, this practice may eventually be deemed suitable for inclusion in California organic strawberry production.

ADDITIONAL INFORMATION

The *UC IPM Pest Management Guidelines for Strawberry* (UC ANR Publication 3468, online at www.ipm.ucdavis.edu) provides additional information on these soilborne pathogens and pests as well as on other threats to strawberry production. The dedicated section on organically acceptable management methods in the guidelines is regularly updated. You can also consult *Integrated Pest Management for Strawberries, Second Edition* (UC ANR Publication 3351), for more information on production and disease management or the *Compendium of Strawberry Diseases, Second Edition* (American Phytopathological Society Press), a comprehensive resource for diagnosis and treatment of strawberry diseases.

REFERENCES/RESOURCES

Blevins, R. L., W. W. Frye, and M. S. Smith. 1985. The effects of conservation tillage on soil properties. Pp. 99–110 in F. M. D'Itri (ed.), A systems approach to conservation tillage. Chelsea, Michigan: Lewis Publishers, Inc.

Brown, P. D., and M. J. Morra. 1997. Control of soil-borne plant pests using glucosinolate-containing plants. Advances in Agronomy 61:167–231.

Bull, C. T., K. G. Shetty, and K. V. Subbarao. 2002. Interactions between myxobacteria, plant pathogenic fungi, and biological control agents. Plant Disease 86:889–896.

Chan, M. K. Y., and R. C. Close. 1987. Aphanomyces root rot of peas. 3. Control by the use of cruciferous amendments. New Zealand Journal of Agricultural Resources 30:225–233.

Chew, F. S. 1988. Biological effects of glucosinolates. Pp. 155–181 in H. G. Cutter (ed.), Biologically active natural products: Potential use in agriculture. Washington, D.C.: American Chemical Society.

Conway, K. E. 1996. An overview of the influence of sustainable agricultural systems on plant diseases. Crop Protection 15:223–228.

Cook, R. J. 1984. Root health: Importance and relationship to farming practices. Pp. 111–127 in D. F. Bezdicek and J. E. Power (eds.), Organic farming: Current technology and its role in a sustainable agriculture. Madison, Wisconsin: American Society of Agronomy Special Publication.

———. 1988. Biological control and holistic plant care in agriculture. American Journal of Alternative Agriculture 3:51–62.

———. 1991. Challenges and rewards of sustainable agriculture research and education. Pp. 32–76 in Sustainable agriculture research and education in the field: A proceedings. National Research Council. Washington, D. C.: National Academy Press.

Curl, E. A., and B. Truelove. 1985. The rhizosphere. Advanced series in agricultural sciences. New York: Springer-Verlag.

Evans, G., and A. C. Gleeson. 1973. Observations on the origin and nature of *Verticillium dahliae* colonizing plant roots. Australian Journal of Biological Sciences 26:151–161.

Fenwick, G. R., R. K. Heaney, and W. J. Mullin. 1983. Glucosinolates and their breakdown products in food and food plants. CRC Critical Rev. in Food Science and Nutrition 18:123–201.

Gamliel, A., and J. J. Stapleton. 1993. Effect of chicken compost or ammonium phosphate and solarization on pathogen control, rhizosphere microorganisms, and lettuce growth. Plant Disease 77:886–891.

Gliessman, S. R., M. R. Werner, S. L. Swezey, E. Caswell, J. Cochran, and F. Rosado-May. 1996. Conversion to organic strawberry management changes ecological processes. California Agriculture Journal 50:24–31.

Gunnell, P. S., and W. D. Gubler. 1992. Taxonomy and morphology of *Colletotrichum* species pathogenic to strawberry. Mycologia 84:157–165.

Harrier, L. A., and C. A. Watson. 2004. The potential role of arbuscular mycorrhizal (AM) fungi in the bioprotection of plants against soilborne pathogens in organic and/or other sustainable farming systems. Pest Management Science 60:149–157.

Harris, D. C., and J. R. Yang. 1996. The relationships between the amount of *Verticillium dahliae* in soil and the incidence of strawberry wilt as a basis for disease risk prediction. Plant Pathology 45:106–114.

Hoitink, H. A. J., A. G. Stone, and D. Y. Han. 1997. Suppression of plant diseases by composts. HortScience 32:184–187.

Hornby, D. 1990. Root diseases. Pp. 232–257 in J. M. Lynch (ed.), The rhizosphere. New York: Wiley.

Jordan, V. W. L., B. Sneh, and B. P. Eddy. 1972. Influence of organic soil amendments on *Verticillium dahliae* and on the microbial composition of the strawberry rhizosphere. Annals of Applied Biology 70:139–148.

Katan, J., and J. E. DeVay. 1991. Soil solarization. Boca Raton, Ann Arbor, Boston, London: CRC Press Inc.

Koike, S. T. 2008. Crown rot of strawberry caused by *Macrophomina phaseolina* in California. Plant Disease 92:1253.

Koike, S. T., S. C. Kirkpatrick, and T. R. Gordon. 2009. Fusarium wilt of strawberry caused by *Fusarium oxysporum* in California. Plant Disease 93:1077.

Labeda, D. P. 1986. New actinomycetes of commercial importance. In F. Megusar and M. Gantar (eds.), Proceedings of the fourth international symposium on microbial ecology, Ljubijana, Slovenia. 271–276.

LaMondia, J. A., and S. B. Martin. 1989. The influence of *Pratylenchus penetrans* and temperature on black root rot of strawberry by binucleate *Rhizoctonia* spp. Plant Disease 73:107–110.

Lazzeri, L., and L. M. Manici. 2001. Allelopathic effect of glucosinolate containing plant green manure on *Pythium* sp. and total fungal population in soil. HortScience 37:1283–1289.

Lewis, J. A., and G. C. Papavizas. 1970. Evolution of volatile sulfur-containing compounds from decomposition of crucifers in soil. Soil Biology and Biochemistry 2:239–246.

———. 1971. Effect of sulfur-containing volatile compounds and vapors from cabbage decomposition on *Aphanomyces euteiches*. Phytopathology 61:208–214.

———. 1974. Effect of volatiles from decomposing plant tissues on pigmentation, growth, and survival of *Rhizoctonia solani*. Soil Science 118:156–163.

Lockwood, J. L. 1988. Evolution of concepts associated with soilborne plant pathogens. Annual Review of Phytopathology 26:93–121.

Lynch, J. M. 1982. Interactions between bacteria and plants in the root environment. Pp. 1–23 in M. E. Rhodes-Roberts and F. A. Skinner (eds.), Bacteria and plants. New York: Academic Press.

Maeder P., A. Fließbach, D. Dubois, L. Gunst, P. Fried, and U. Niggli. 2002. Soil fertility and biodiversity in organic farming. Science 296(5573):1694–1697.

Martin, F. N. 2001. Management of pathogens associated with black root rot of strawberry. Abstract 46 in Annual International Research Conference on Methyl Bromide Alternatives and Emissions Reductions. Accessed at www.mbao.org/2001proc/046%20Martin%20FN%20mbao%20-%2001%20abstract-b.pdf.

Martin, F. N., and C. Bull. 2002. Biological approaches for control of root diseases in strawberry. Phytopathology 92:1356–1362.

Menzies, J. G., and R. R. Belanger. 1996. Recent advances in cultural management of diseases of greenhouse crops. Canadian Journal of Plant Pathology 18:186–193.

Milholland, R. D. 1994. A monograph of *Phytophthora fragariae* and the red stele disease of strawberry. North Carolina Agricultural Research Service Technical Bulletin 306.

Nederman, G. C. 1991. Effects of crop residue and tillage practices and crop production. Pp. 23–24 in Cover crops for clean water. Ankeny, Iowa: Soil and Water Conservation Society.

Njoroge, S. M. C., Z. Kabir, F. N. Martin, S. T. Koike, and K. V. Subbarao. 2009. Comparison of crop rotation for Verticillium wilt management and effect on *Pythium* species in conventional and organic strawberry production. Plant Disease 93:519–527.

Olivier, C., S. F. Vaughn, E. S. G. Mizubuti, and R. Loria. 1999. Variation in allyl isothiocyanate production within *Brassica* species and correlation with fungicidal activity. Journal of Chemical Ecology 25:2687–2701.

Pettitt, T. R., and G. F. Pegg. 1994. Sources of crown rot (*Phytophthora cactorum*) infection in strawberry and the effect of cold storage on susceptibility to the disease. Annals of Applied Biology 125:279–292.

Roberson, E. B., S. Sarig, and M. K. Firestone. 1991. Cover crop management of polysaccharide-mediated aggregation in an orchard soil. Soil Science Society of America Journal 55:734–739.

Rovira, A. D., L. F. Elliot, and R. J. Cook. 1990. The impact of cropping systems on rhizosphere organisms affecting plant health. Pp. 389–435 in J. M. Lynch (ed.), The rhizosphere. New York: Wiley.

Stapleton, J. J. 1994. Solarization as a framework for alternative soil disinfestation strategies in the interior valleys of California. In Proceedings of the 1994 Annual International Research Conference on Methyl Bromide Alternatives and Emissions Reductions. Kissimmee, Florida.

Subbarao, K. V., and J. C. Hubbard. 1996. Interactive effects of broccoli residue and temperature on *Verticillium dahliae* microsclerotia in soil and on wilt in cauliflower. Phytopathology 86:1303–1309.

Subbarao, K. V., J. C. Hubbard, and S. T. Koike. 1999. Evaluation of broccoli residue incorporation into field soil for Verticillium wilt control in cauliflower. Plant Disease 83:124–129.

Subbarao, K. V., Z. Kabir, F. N. Martin, and S. T. Koike. 2007. Management of soilborne diseases in strawberry using vegetable rotations. Plant Disease 91:964–972.

Umaerus, V. 1992. Crop rotation in relation to crop protection. Netherlands Journal of Plant Pathology 98:241–249.

van Bruggen, A. H. C. 1995. Plant disease severity in high-input compared to reduced-input and organic farming systems. Plant Disease 79:976–984.

Vigo, C., J. R. Norman, and J. E. Hooker. 2000. Biocontrol of the pathogen *Phytophthora parasitica* by arbuscular mycorrhizal fungi is a consequence of effects on infection loci. Plant Pathology 49:509–514.

Wilhelm, S., P. Nelson, H. Thomas, and H. Johnson. 1972. Pathology of strawberry root rot caused by *Ceratobasidium* species. Phytopathology 62:700–705.

Williams, W. A. 1966. Management of non-leguminous green manures and crop residues to improve the infiltration rate of an irrigated soil. Soil Science Society of America Proceedings 30:631–634.

Wing, K. B., M. P. Pritts, and W. F. Wilcox. 1994. Strawberry black root rot: a review. Advances in Strawberry Research 13:13–19.

Winterbottom, C., F. Westerlund, J. Mircetich, L. Galper, and N. Welch. 1996. Evaluation of relative resistance of different strawberry cultivars to *Phytophthora* and *Verticillium dahliae* as a potential alternative to methyl bromide. Proceedings from Annual International Research Conference on Methyl Bromide Alternatives and Emissions Reductions, Nov. 4–6, 1996, Orlando, Florida. 36:1–5.

Xiao, C. L., K. V. Subbarao, K. F. Schulbach, and S. T. Koike. 1998. Effects of crop rotation and irrigation on *Verticillium dahliae* microsclerotia in soil and wilt in cauliflower. Phytopathology 88:1046–1055.

Yuen, G. Y., M. Schroth, A. Weinhold, and J. Hancock. 1991. Effects of soil fumigation with methyl bromide and chloropicrin on root health and yield of strawberry. Plant Disease 75:416–420.

PHOTO: ISTOCKPHOTO.COM

10 Arthropod Management in Organic Strawberries

SURENDRA DARA, JIANLONG BI, AND MARK BOLDA

LYGUS BUGS

Lygus bugs are important pests of strawberries in coastal California from the Central Coast region to Ventura County. The western tarnished plant bug (*Lygus hesperus*), tarnished plant bug (*L. lineolaris*), and pale legume bug (*L. elisus*) are the common species found in this area. Lygus bug feeding causes significant reductions in the yield and quality of strawberry fruit. The fruit distorted from lygus bug damage become unfit for fresh market sales. Lygus infestation is more severe when fruit is continuously present in the fields.

Biology

The lygus bug life cycle includes eggs, nymphs, and adults (figure 10.1). Adults are about 6 mm long and yellowish green to brownish. Depending on temperature conditions, the egg period, early nymphal stage (first to third instars), and late nymphal stage (fourth to fifth instars) can last for about 7 days each. Adults may take as long as 9 days before they start laying eggs, and egg laying can continue for 21 days.

On strawberries, the majority of the eggs are laid in the inflorescences rather than the leaves. Eggs are inserted into the plant tissue. Newly emerged nymphs are small and pale. They are easily confused with aphids but can be distinguished by their fast movement. Older nymphs can be recognized by five black spots on their back (four on the thorax and one in the middle of the abdomen).

Damage

Several flowering weeds and legumes, like alfalfa, are preferred by lygus over strawberries, but infestations in strawberries can still cause significant damage (figure 10.2). Lygus bugs feed by puncturing the plant tissue and sucking out its juices, which causes damage to achenes (the strawberry seeds). Portions of the fruit where the achenes are damaged fail to develop normally, resulting in a fruit deformation known as cat-facing. Severe damage results from feeding during flowering. Fruit deformation due to lygus feeding has a similar appearance to damage from low-temperature injury, except that with lygus damage the achenes in the affected area are of a normal size. Low-temperature injury results in achenes in the affected areas that are smaller than they are on the rest of the fruit.

Figure 10.1. Life stages of lygus bug (western tarnished plant bug, *Lygus hesperus*): (A) lygus first instar nymph (right) compared to the size of an aphid (left); (B) first instar (see spot on the abdominal segment); (C) second instar (only one spot on the abdominal segment); (D) third instar (still only one spot on the abdominal segment); (E) fourth instar (five spots prominent, wing pads visible); (F) fourth instar, side view; (G) fourth instar (four spots on thorax and one on abdomen); (H) fifth instar developing wing pads; (I) *Lygus hesperus* adult. *Photos* by S. Dara and R. Cooper.

Figure 10.2. Strawberries deformed due to lygus bug feeding damage. *Photos* by S. Dara and J. K. Clark.

Sampling

A common sampling technique is to divide the field into four blocks and sample four 200-foot lengths along the row at random locations within each block. Sample 10 plants per 200-foot row by beating the plant and collecting lygus bugs into a tray placed under the plant. The treatment threshold is one lygus nymph per 20 plants. If an insect vacuum is used for sampling, the threshold is one lygus bug per 10 plants.

Sampling for lygus in alternate hosts in the vicinity of the strawberry field can also give you information that is helpful when you are making treatment decisions. Depending on the type of host plant, sampling can be accomplished with a sweep net, beating tray, insect vacuum, or even visual inspection. Knowledge of lygus populations outside of the field can give you a good indication of their potential for movement into as well as out of the strawberry field.

Management Options

- *Weed management.* Manage weeds in winter. Once lygus adults emerge, they will migrate to strawberries when the weeds senesce or are removed. Mow or disk weeds and alternate hosts while the lygus are in the nymphal stage. This will remove a major source of infestation that would otherwise spread to strawberries.
- *Trap crops.* Grow flowering host plants near strawberries to attract lygus and then manage lygus populations on these hosts; this requires careful management.
- *Vacuuming.* Some organic strawberry growers use an insect vacuum to control the lygus bugs in strawberry fields. They vacuum strawberry plants several times a week when the lygus population is high. Strips of trap crops like alfalfa are grown along the edges or at regular intervals within the strawberry field and vacuumed with tractor-mounted machines. The downside to vacuuming is that it can also remove beneficial arthropods.
- *Cultivars.* Observations suggest that certain cultivars with abundant foliage appear to harbor smaller lygus populations than cultivars with abundant inflorescences. Lygus bugs' preference for depositing their eggs in flowers and fruit rather than in leaves is probably a contributing factor toward this difference. If you choose varieties that are less attractive to lygus, you can help reduce the pest load.
- *Biological control.* Generally speaking, biological control of lygus is poor, especially in fields where broad-spectrum insecticides are used. However, in cases where selective sprays are used, several naturally occurring predators such as big-eyed bugs (*Geocoris* spp.), damsel bugs (*Nabis* spp.), minute pirate bug (*Orius tristicolor*), and several species of spiders feed on lygus (figure 10.3). A mymarid wasp (*Anaphes iole*) that attacks lygus eggs is commercially available. It can reduce lygus numbers but will not bring them below threshold levels. These wasps are less effective against eggs that are in the strawberry inflorescence (where more eggs are present) than against eggs that are in foliage. Recent studies have also documented some effectiveness from two species of braconid parasitoids (*Peristenus* spp.) that attack lygus nymphs.

Figure 10.3. Predaceous insects that attack lygus bug nymphal stages: (A) bigeyed bug, (B) damsel bug adult and nymph, and (C) minute pirate bug. *Photos* by J. K. Clark.

Figure 10.4. Lygus bug adult killed by the insect-pathogenic fungus *Beauveria bassiana*. The fungus emerges from the dead insect and produces spores that infect other insects on contact. *Photo* by S. Dara

- *Microbial control.* Various entomopathogenic fungi including *Beauveria bassiana* (figure 10.4), *Metarhizium anisopliae, Hirsutella* spp., *Isaria* spp., and *Verticillium* spp. are pathogenic to nymphal and adult stages of lygus. Limited studies on the use of entomopathogenic fungi to manage strawberry pests including lygus have shown promising results. Isolates of some of these fungi are commercially available, and a *B. bassiana*–based product is registered for use on organic strawberries. Mild temperatures and foggy conditions in California's Central Coast area are congenial for entomopathogenic fungi. The strawberry plant, being close to the ground and spread out, is also ideally suited to protect the fungal inoculum under its canopy.
- *Other control products.* Some organically approved insecticides are available for lygus control, but it is important that you verify their efficacy and potential for phytotoxicity before applying them. Insecticidal soap (M-Pede) contains potassium salts of fatty acids that abrade the insect cuticle. Petroleum (SuffOil-X) and paraffinic (Organic JMS Stylet Oil) oils act by suffocating the insects.
- *Second-year or cutback fields.* Second-year or cutback strawberry fields extend the availability of the lygus host beyond the normal growing season. Existing populations combine with new populations that migrate from other fields to increase lygus numbers. Such fields also serve as a source of early infestation for the newly planted fields. By avoiding the extension of a strawberry field to a second year of production, you can reduce your risk of lygus infestations.
- *Location of the field.* Planting strawberries in fields far from common sources of infestation is another important measure. This can reduce or delay lygus bug infestation.

SPIDER MITES

Spider mites are important pests of strawberries in California. These tiny arthropods are similar to insects in that they have jointed legs, but biologically they are more closely related to spiders and ticks, as can be seen by their four pairs of legs and two body segments. The most important spider mite species for California strawberry production is the twospotted spider mite *(Tetranychus urticae)*, which is a major pest in coastal areas. Strawberry spider mite *(T. turkestani)*, which resembles *T. urticae*, may also be present among spider mite populations but prefers warmer climates. The two species look very much alike and are yellow or greenish with dark spots on either side of the body. Microscopic examination of the male genitalia is necessary to distinguish one species from the other.

Heavy infestations of another spider mite known as Lewis mite or Lewis spider mite *(Eotetranychus lewisi)* have been reported recently in Ventura County. This species can be confused with twospotted spider mite, but adult Lewis mite females are smaller than adult twospotted females. Lewis mites have several small spots on the body, while twospotted spider mites have a single dark spot on either side of the body.

Carmine spider mite *(T. cinnabarinus)* is bright red and can be found at low densities on strawberries. Spider mite feeding reduces the number, size, and quality of strawberries.

Biology

Mites are not insects. They are more closely related to ticks and spiders. While an insect has three body parts (head, thorax, and abdomen) and three pairs of legs, a mite's body is divided into two major parts, and it has four pairs of legs.

The life cycle of a spider mite includes egg, larva, protonymph, deutonymph, and adult stages (figure 10.5). Eggs of the twospotted spider mite are round and clear; those of the carmine mite are yellowish or orange. The larvae have three pairs of legs, but nymphs and adults have four pairs. The duration of the mites' life cycle depends on temperature. Under ideal conditions, spider mites can complete a full generation in as little as 1 week, though generation times of about 2 weeks are more common during much of the strawberry growing season. Adult females are larger than males and can measure as long as 0.4 mm. The ratio of male to female mites varies depending on a number of factors, but it is usually 1 male to every 2 or 3 females.

Favorable conditions in the strawberry growing regions of the California coast mean that active spider mites can be observed there throughout the year. They overwinter as dormant adult females in cooler periods and areas and remain active through the winter where conditions are milder. Overwintering twospotted spider mite females are reddish orange and can be confused with other mites.

Figure 10.5. Twospotted spider mite (A) egg and protonymph, (B) adult male and (C) female, (D) strawberry spider mite adult, (E) carmine spider mite protonymph and adult, and (F) Lewis mite. *Photos* by S. Dara, R. Cooper, and J. K. Clark.

Damage

Spider mites feed on the undersides of leaves and cause significant reductions in strawberry yields. Damage starts as a stippling of yellow spots on leaves and then advances to scarring and bronzing of leaves (figure 10.6). Webbing is visible on infested leaves; hence the name spider mite. Severe infestations can stunt plant growth and kill the plant. Fall-planted strawberries are very sensitive to mite damage for the first 2 to 5 months after transplanting. More than one mite per leaflet during this period can result in a measurable loss in yield. Plants are less sensitive to mite damage after initial berry set, but substantial losses can occur if there are 15 to 20 mites per mid-tier leaflet during this period. Highest population levels in coastal strawberry production are usually seen before the peak of the first crop, and populations decline in June. For summer-planted strawberries, increases in spider mite populations can also occur in the fall.

Risk Factors

An early buildup of mite populations will result in season-long damage. Fall plantings are generally more susceptible than summer plantings. Day-neutral cultivars are also more susceptible than short-day cultivars. Fertilizers influence the carbohydrate and phenol content of the plants, and high carbohydrate or leaf nitrogen levels promote the oviposition, early development, and fecundity of mites. Phenols are defense chemicals that occur naturally in plants, and their reduction makes for a less antagonistic situation for mites. Proximity to second-year plantings with mite infestation, the presence of infested fields upwind, inadequate chilling of day-neutral varieties, dusty conditions, and water stress are other factors that contribute to mite problems.

Sampling

Depending on the region and the time of the year, monitor for mites every week or every other week. When populations are low or when you are monitoring a small acreage, check for the presence of mites by randomly sampling 10 leaflets per acre. When infestation is high, 5 to 10 leaflets per acre

Figure 10.6. (A) Stunted strawberry plant with discolored leaves from spider mite damage. Dead or dying plant tissue from mite feeding results in (B) brown spots on the lower side and (C) yellowish or brownish spots on the upper side of the leaves. *Photos* by S. Dara.

can be randomly sampled and counted for mites. Economic thresholds during the first 4 months after planting are 5 mites per mid-tier leaflet for fall plantings and 10 mites per mid-tier leaflet for summer plantings.

Management Options

- *Plant health.* Young and healthy plants can withstand mite damage better than older plants. Promote vigorous plant growth through adequate chilling, water, and nutritional management.
- *Fertilization.* Avoid excessive nitrogen fertilization as it can increase mite infestations.
- *Dusty conditions.* To prevent dust buildup on the plants, use low fences, drive slowly, and water down roadways.
- *Biological control.* Conserve or release natural enemies such as predatory mites. *Phytoseiulus persimilis* is a specialist predator and is effective early in the season. *Neoseiulus californicus,* which tolerates high temperatures and a wide range of humidity, is a generalist predator. It is naturally occurring and is the predominant species later in the season. Both species are commercially available for field release and are most successful in fields where no pesticides (or only selective pesticides) are used. Several predatory insects also provide biological control of spider mites, including the minute pirate bug *(Orius tristicolor),* a small, black lady beetle (*Stethorus* spp.), a small, black rove beetle *(Oligota oviformis),* big-eyed bugs (*Geocoris* spp.), brown lacewings (*Hemerobius* spp.), green lacewings (*Chrysopa* spp.), sixspotted thrips *(Scolothrips sexmaculatus),* and damsel bugs (*Nabis* spp.).
- *Microbial control.* Many entomophagous fungi such as *Beauveria bassiana, Metarhizium anisopliae,* and species of *Hirsutella* and *Isaria* are pathogenic to many mites including those that attack strawberries, but information on their use in strawberries is limited.
- *Other control products.* Cinnamaldehyde, insecticidal soaps, petroleum-based oils, and botanical oils are some of the products available in the marketplace that have different modes of action for mite control. These can be used in late winter before predatory mites are released. Cinnamaldehyde is thought to interfere with glucose uptake and utilization. Insecticidal soaps have fatty acids that abrade mite skin. Paraffinic and petroleum oils suffocate the eggs and other life stages of mites. Rosemary oil is believed to repel mites and to affect oviposition behavior. However, because the labels of some of these products change frequently and because certain soap- and oil-based products can cause phytotoxicity, it is important to evaluate their safety and efficacy on localized areas before you spray them on an entire field.

PREDATORY MITES

Different members of the Phytoseiidae family of mites feed on phytophagous mites like those that attack strawberries. A specialist predatory mite, *Phytoseiulus persimilis,* and a generalist predatory mite, *Neoseiulus californicus,* are important among them (figure 10.7). They feed on all life stages of spider mites. Although *P. persimilis* is a voracious feeder, it is a specialist predator and will not persist when no prey mites are available. It needs spider mites for its growth and reproduction, but it can also feed on young thrips. Alternatively, *N. californicus* can survive and reproduce by feeding on a variety of phytophagous mites, thrips, and other small insects. Both of these species can feed on each other and develop to maturity, but cannibalism is greater in *P. persimilis* in the absence of prey.

Phytoseiulus persimilis

Adult *Phytoseiulus persimilis* are reddish orange and teardrop-shaped, and move rapidly with their long legs. They are slightly larger than spider mites, and their eggs are round and bigger than those of spider mites. The larvae do not feed at all; only nymphs and adults feed on their prey. The mites' life cycle can take anywhere from 5 to 25 days, depending on temperature. Adults feed on 5 to 20 prey per day and can live as long as 50 days. Long setae (hairs) on their dorsal side are thought to help them navigate through the webbing of the spider mites.

Figure 10.7. Predatory phytoseiid mites (A) *Phytoseiulus persimilis* and (B) *Neoseiulus (Amblysius) californicus,* which are major predators of spiders mites on strawberries. *Photos* by J. K. Clark.

Spider mites are the preferred prey of *P. persimilis* over other prey species. The host plant of the prey does not influence *P. persimilis,* which is well adapted to survive in patchy prey populations. A preference for cooler temperatures and a need for high relative humidity make *P. persimilis* effective against spider mites early in the strawberry season when temperatures are mild.

In the absence of prey, *P. persimilis* will starve or cannibalize and will eventually disappear. It is also sensitive to most miticides and can be killed by miticide applications. Certain pesticides and fungicides repel *P. persimilis* from the eggs of prey and so affect their predation.

When spider mite populations are limited to localized areas, 40,000 to 60,000 *P. persimilis* can be released per acre. When spider mites are well distributed but present in low numbers, release only 30,000 per acre. If spider mite populations are on the increase, 100,000 or more *P. persimilis* may be released per acre. It is important to remember that this predator is effective early in the season when temperatures are cool.

Neoseiulus californicus

Neoseiulus californicus adults are oval and translucent with shades of pale orange or peach. Adult females are about 0.1 mm in length. The pale white football-shaped eggs are visible only with a hand lens. Larvae can develop into protonymphs without feeding. Both nymphal and adult stages actively feed on their prey. Rate of development depends on temperature; the mites can take 4 to 12 days to complete their life cycle and reach adulthood. The adult stage lasts about 20 days.

While *N. californicus* can develop on a variety of prey species, their growth and development are rapid when they feed on twospotted spider mite. Predatory mites like *N. californicus* are able to disperse, are well adapted to changing environmental conditions like those in agricultural systems, and are able to tolerate certain miticides.

Neoseiulus californicus prefers warmer climates and can withstand a wide range of relative humidity. It can also survive brief periods of cold, can fare well in the warmer times of year, and is usually the dominant species later in the strawberry season.

One female per plant or about 30,000 females per acre is the recommended rate of release for *N. californicus.*

CYCLAMEN MITE

Cyclamen mite, *Phytonemus pallidus*, has been found increasingly in organic strawberry fields (figure 10.8). Cyclamen mite is much smaller than spider mites and causes no webbing. It belongs to the family Tarsonemidae and is so small it can only be seen with a hand lens. Cyclamen mites are normally found at the midvein of unopened leaves and under the calyx of flower buds. The activities of this pest lead to a crinkling of the new strawberry leaves as they emerge from the crown. Severe infestations can cause total stunting of plants and substantially reduced yields (figure 10.9).

Biological control of cyclamen mite in commercial production fields typically cannot keep populations below economically damaging levels. For this reason, the best place to manage cyclamen mites is at the nursery where the plants are produced. Growers should take care to obtain plant stock that has not been infested in the first place or that has undergone hot water treatments. Details regarding hot water treatments can be found in the *UC IPM Pest Management Guidelines for Strawberries* (UC ANR Publication 3468). Cyclamen mite can be spread throughout a field or from field to field by pickers and farm equipment, or it can be spread by birds and insects. For this reason, it is best to identify plants that are visibly infested with cyclamen mites and manually remove them from the field.

Figure 10.8. Cyclamen mite eggs and adults. *Photo* by J. K. Clark.

Figure 10.9. Severe cyclamen mite infestation causes stunted and crinkled leaves in the center of the strawberry plant. *Photo* by J. K. Clark.

GREENHOUSE WHITEFLY

The greenhouse whitefly (*Trialeurodes vaporariorum*) is an important pest of strawberries in the coastal production regions of California, especially near Oxnard and Ventura. Whiteflies can reduce strawberry yields and cause further damage by excreting honeydew, a liquid that leads to the growth of sooty mold. Both honeydew and sooty mold are contaminants that lower the quality of strawberry fruit.

Biology

Whitefly adults are about 1 mm long and have four membranous wings coated with white, powdery wax. The eggs are elongated, range from pale yellowish green to brown, and are attached to the underside leaf surface by a short pedicel. Eggs turn black before hatching. The immatures are oval, thin, flat, and often semitransparent. The pupae are recognizable by the long, waxy filaments attached to their body. Adults are white and hold their wings flat over their back.

The greenhouse whitefly goes through six stages during its development: egg, then first, second, third, and fourth instar nymphs, and finally adult (figure 10.10). The whitefly does not have a true pupal stage, but the fourth instar nymph is often called a pupa or red-eye nymph due to the presence of two prominent red eyes. Only adults and newly hatched first instar nymphs are mobile. When temperatures are warm, whiteflies can complete their life cycle in about 3 weeks.

The first instar nymphs that hatch out of the eggs are known as crawlers. These crawlers disperse on the host plant until they find a suitable feeding site. They then become immobile until they develop into adults.

The adult whitefly emerges from a T-shaped slit in the fourth instar. It may take several hours after emergence for the whitefly's wings to dry before it is able to fly. Whiteflies breed and develop in continuous cycles as long as temperatures are suitable. Adults make relatively short flights from plant to plant or from field to field, but whiteflies may be carried great distances by wind or may be transported on host plants.

The greenhouse whitefly has a large host range and can be destructive to many crop species in strawberry production areas.

Figure 10.10. Greenhouse whitefly life stages: (A) eggs, (B) nymphs, and (C) an adult next to empty pupal cases. *Photo* by J. K. Clark.

Damage

Whiteflies feed on the sap of their host plant. Large populations of the greenhouse whitefly can ingest sufficient quantities of plant sap to cause yield reductions and plant stunting. In addition, this pest produces large amounts of honeydew and associated black sooty mold, which decreases the number of marketable strawberries. Further, this insect transmits plant virus diseases such as strawberry pallidosis–associated virus and beet pseudo yellows virus, which can hasten decline in strawberry plantings.

Monitoring

Yellow sticky traps and leaf counts are the standard monitoring methods for whiteflies. Yellow sticky traps are most commonly used to monitor whitefly movement into fields. For every 10 acres of planting, place one sticky trap at the edge of the strawberry field on a vertical stake just above the plant canopy. Monitor the traps weekly for migration into the field. Once populations become established, switch to foliar inspections that provide a better estimate of whitefly densities. Inspect 20 terminal leaflets in each quarter of the field for numbers of adults and check 20 mid-aged leaflets for numbers of nymphs. Research shows that a population of 10 to 20 adults per leaflet will reduce the soluble sugars in the berries. A March population of 10 adult whiteflies per leaflet has been shown to decrease strawberry yields by 10 percent, and a population of 50 per leaflet has been shown to reduce yields by almost 25 percent.

Management Options

- *Cultural control.* Greenhouse whitefly populations move back and forth among various hosts in intensive farming areas. It is important to break the pest's continuous development cycle by not growing the hosts side by side and to adopt proper sanitation practices after crop harvest.
- *Biological control.* Natural enemies like big-eyed bugs (*Geocoris* spp.), minute pirate bugs (*Orius* spp.), lacewing larvae (*Chrysopa* spp. and *Chrysoperla* spp.), and several native parasitic wasps such as *Encarsia pergandiella* keep green-

house whitefly populations below the levels that would cause damage. Although augmentative releases of parasites have been successful in greenhouses, releases of commercially available species such as *Encarsia formosa* and *Eretmocerus eremicus* have not been successful in California strawberry fields where the greenhouse whitefly is a problem. One UC Davis study in the Watsonville area showed that early-season releases of the predatory mite *Amblyseius swirskii* suppressed whitefly populations for as long as 2 months.

- *Microbial control.* There are several genera of entomopathogenic fungi that cause natural infections in whiteflies. Important pathogens among them are *Beauveria bassiana, Metarhizium anisopliae, Isaria fumasoroseus,* and *Verticillium lecanii,* and some of these are available commercially. The efficacy of these pathogens on whitefly control in strawberry fields, however, has not been established.
- *Chemical control.* Narrow-range oil, azadirachtin, and insecticidal soap can be sprayed against low to moderate populations of greenhouse whitefly.

WESTERN FLOWER THRIPS

Western flower thrips, *Frankliniella occidentalis,* is native to western North America and is widespread throughout strawberry production regions in California.

Biology

Western flower thrips is a long, narrow insect with a body length of less than 2 mm (figure 10.11). Nymphs are white or yellowish and adults range from yellow to dark brown. Adults have wings that have fringes or microscopic hairs.

The female thrips lays its eggs inside leaves, flower buds, or petals. Eggs hatch into immature thrips that develop from first to second instar larvae. These larvae feed on flower buds or terminal foliage. Toward the end of the second instar, the thrips stop feeding and drop to the soil or leaf litter where they become prepupae and then pupae. Larval stages are wingless, and wing buds become visible in the prepupae and pupae. Winged adults then emerge from the pupae. Developmental times vary depending on temperature and the host plant species. At 59°F the life cycle is completed in about 2 weeks, while at 86°F the life cycle can be completed in 4.3 days. In spring, thrips populations build up on nearby flowering weeds. When the weeds are cut, stop flowering, or dry up, the thrips migrate to strawberries.

Damage

Thrips use their mouthparts to pierce plant cells and suck out their contents. The damaged plant cells collapse. Damage in strawberries is described as fruit bronzing, a tan or brown discoloration, and sometimes leads to desiccation and cracking of the fruit surface (figure 10.12).

There are three types of strawberry bronzing damage. Type I bronzing is caused by thrips feeding,

Figure 10.11. Western flower thrips adult. *Photo* by J. K. Clark.

Figure 10.12. Browning of the fruit around the cap due to thrips damage, known as Type I bronzing. *Photo* by J. K. Clark.

while Type II bronzing is the result of spray burn from agricultural chemicals, including sulfur. In both cases, bronzing is visible in localized areas of the fruit surface that were directly exposed to thrips feeding or chemicals. Type III bronzing, which covers the entire fruit surface, occurs under conditions of elevated temperature and solar radiation.

Monitoring

Shake strawberry flowers over a flat surface to count the thrips. Consider treatment when thrips populations reach 10 per flower. A better way to sample for thrips is to place randomly collected flowers into glass vials with several drops of ethyl acetate or methyl isobutyl ketone. After half an hour, shake the vial, and remove the flower, and count the thrips against a piece of black paper. Because more thrips will be found with this method, the treatment threshold is higher than for shaken flowers.

Management Options

- *Biological control.* Minute pirate bugs (*Orius* spp.), the predominant predator of thrips, are available commercially, though no formal recommendations have been developed for their release rates or timings.
- *Microbial control.* Several entomopathogenic fungi such as *Beauveria bassiana, Metarhizium anisopliae, Isaria fumasoroseus,* and *Verticillium lecanii* can cause natural infections in thrips. The efficacy of these pathogens for thrips control in strawberry fields has not been tested.
- *Chemical control.* Sprays of spinosad (Entrust) and pyrethrin (PyGanic) are organically acceptable for the control of western flower thrips.

LEPIDOPTEROUS PESTS

Leafrollers

Several leafroller species, including the orange tortrix *(Argyrotaenia citrana),* garden tortrix *(Ptycholoma peritana),* strawberry leafroller *(Ancylus comptana fragariae),* and light brown apple moth *(Epiphyas postvittana),* occur in strawberries (figure 10.13). Normally well under 1 inch long, leafrollers roll leaves with abundant webbing and are generally green to brown. The larvae wriggle about vigorously when touched. Adult moths have a distinctive bell shape when the wings are closed, are brown or mottled brown, and are about 13 mm long. Leafrollers rarely occur in sufficient numbers or cause sufficient damage to warrant control measures. However, the light brown apple moth may require treatments due to government regulations related to this pest, which is relatively new to California.

A number of organic treatment options are available for leafroller control in organic strawberries. Products containing *Bacillus thuringiensis* or spinosad can be effective. There are many predators and parasitoids of leafrollers, and most often they will keep populations adequately managed. Avoiding second-year strawberry plantings can be very helpful in keeping leafroller density down.

Cutworms

Cutworms, mostly *Agrotis ipsilon,* cause occasional problems in strawberries when they feed on fruit and young plants. The cutworm larvae are a little bit larger than other lepidopterous larvae, ranging from 25 to 30 mm long (figure 10.14). Cutworms have a greasy black to gray appearance and are mostly found in their distinctive C-shaped resting position. Feeding mostly at night, they can be found during the day at the base of the plant and underneath debris or the plastic tarp.

To manage cutworms, follow good sanitation practices. Annual plant culture that reduces the amount of weeds and debris around the plant, along with thorough pruning of second-year plantings, will effectively limit cutworm populations. In cases where good sanitation is not enough in itself, cutworms can be managed with *B. thuringiensis,* which is most effective against smaller stages of the worms, especially when applied in the evening to minimize the photodegradation of the active ingredient that results from exposure to sunlight.

Figure 10.13. Larvae of (A) orange tortrix, (B) garden tortrix, and (C) light brown apple moth. *Photos* by J. K. Clark.

Figure 10.14. Larvae of (A) black (or greasy) cutworm and (B) variegated cutworm. *Photos* by J. K. Clark.

VINEGAR FLIES

Drosophila spp. are small brown to tan flies that are attracted to fermenting fruit. Adults are small flies (about 3 mm) with red eyes, a pale brown thorax, and an abdomen with black stripes (figure 10.15). Eggs and larvae are a contamination problem for very ripe or overripe fruit, and adult flies can be a nuisance when numerous. The recently introduced spotted wing drosophila, *Drosophila suzukii*, is a destructive vinegar fly that can infest sound fruit prior to harvest. The male spotted wing drosophila is easy to distinguish from other vinegar flies because of the black spot it has on the back of each wing. Females are harder to distinguish and require examination of the ovipositor under magnification. Spotted wing drosophila ovipositors are serrated and larger than those of other *Drosophila* species.

Little is currently known about the life cycle of spotted wing drosophila in California; like other vinegar flies, it appears to have a short life cycle (from 1 week to a few, depending on temperature) and may go through as many as 10 generations per year. This rapid developmental rate allows it to quickly develop large populations that can inflict severe damage to a crop. Spotted wing drosophila is also highly mobile within localized areas. Strawberry fields in close proximity to other preferred hosts, such as caneberries, are at greater risk for damage than fields that are more isolated.

Adult flies may be captured throughout much of the year, but they are most active at temperatures of 68°F. Activity declines at temperatures above 86°F.

A successful program for the management of vinegar flies in organic strawberries—including the spotted wing drosophila—should focus on control of the flies before they can lay eggs and reduction in the availability of breeding sites. Adult flies can be controlled using broad-spectrum or spinosad-based insecticide baits or sprays. To reduce breeding sites, practice good field sanitation practices that remove rotten fruit from the field. Since vinegar flies are highly mobile, it is important to attempt areawide management programs that will prevent a single, heavily infested field from re-infesting other areas that have received more consistent treatment.

Figure 10.15. (A) Female and (B) male spotted wing drosophila adults. *Photos* by Martin Hauser, California Department of Food and Agriculture.

REFERENCES/RESOURCES

Mueller, S. C., C. G. Summers, and P. B. Goodell. 2003. A field key to lygus species of the central San Joaquin Valley. Oakland: University of California Division of Agriculture and Natural Resources. Publication 8104. Contains information on the identification of lygus species found in the Central Valley of California. Accessed at http://anrcatalog.ucdavis.edu.

UC IPM (University of California Statewide Integrated Pest Management Project). 2008. Integrated pest management for strawberries, Second edition. Oakland: University of California Division of Agriculture and Natural Resources. Publication 3351. Contains details regarding the identification, biology, and management of arthropod pests of strawberries. Available for purchase at http://anrcatalog.ucdavis.edu or at local UC Cooperative Extension offices.

———. 2010. UC IPM pest management guidelines for strawberries. Oakland: University of California Division of Agriculture and Natural Resources. Publication 3468. Contains brief summaries of arthropod pests in strawberries, how to sample for them, and how to manage them, including recommendations for pesticides that are available for use if needed. Accessed at http://anrcatalog.ucdavis.edu.

HARVEST AND POSTHARVEST CONSIDERATIONS

PHOTO: KATHY KEATLEY GARVEY

HARVEST AND POSTHARVEST CONSIDERATIONS

11 Postharvest Handling for Organic Strawberries

ELIZABETH MITCHAM

Strawberries are among the most perishable of fresh fruits, yet they are successfully marketed worldwide in increasing volumes. California strawberry production begins in late winter in the southern part of the state and continues through late fall in the north. Fresh market is the primary outlet, followed by freezing and jam manufacturing markets.

The primary market for fresh strawberries extends across the United States and into southern Canada. Some fruit are exported overseas by air to markets such as Japan and Australia. Our ability to ship strawberries to distant markets depends on the success of preharvest decay control strategies and postharvest temperature management. Most domestic shipments move by highway truck. Surface-transported strawberries must have a 5- to 7-day market life when shipped to eastern U.S. cities.

HARVESTING AND FIELD HANDLING

Strawberry growers and harvesters play a key role in determining the quality and deterioration rate of fruit throughout the marketing and distribution chain. Important factors in the field include

- preharvest disease control and field sanitation
- maturity selection
- avoidance of fruit injury while harvesting fruit and packing it into crates
- grading to eliminate injured, diseased, and defective fruit
- protection from warming
- prompt movement of fruit from field to cooler

Harvesting should be repeated as frequently as is necessary to avoid picking overmature fruit. Any overripe berries should be diverted to roadside markets or to processing, or they should be discarded. Fruit color should be within a fairly narrow range at harvest, normally at least three-fourths red color, so that all harvested berries will perform and respond to handling conditions in a similar manner. The riper the berries are at harvest, the sweeter and more flavorful the fruit will be. Strawberry flavor does not improve after harvest, so fruit should be harvested at the ripest stage that can be safely transported and marketed.

Fruit should be sorted carefully to remove any that have even small decay lesions or injuries (cuts, finger bruises, etc.), and care should be taken to avoid causing any wounds to the fruit during the harvest and packing operations as these will lead to more rapid deterioration and decay. Decay can spread from damaged berries to sound berries in the same basket. Harvesting, grading, and packing can be done simultaneously by pickers in the field, thereby reducing the number of handling steps and the potential for fruit injury (Mitchell et al. 1996). By placing the harvest crates on a picking cart or stand, pickers can keep them off the ground and facilitate easy packing during harvest (figure 11.1). In a small-scale operation, you can pick berries into small containers and transport them to a separate, shaded area for packing. In any case, it is critical that you handle the fruit gently and expeditiously.

Strawberries can be packed for shipment in corrugated fiberboard crates that can hold a variety of types of baskets. Clear, thermoformed plastic clamshell baskets in quart and pint sizes are the most common (figure 11.2). These baskets can be placed directly on retail display without repackaging the fruit or adding a cover. Open-mesh plastic baskets (figure 11.3) are generally used for food service customers. A crate holds 4 to 5 kilograms (9 to 11 lb) of fruit. The plastic baskets can be capped or uncapped, but the crate itself is not lidded for subsequent handling. The critical factor for strawberry packaging is that it must protect the berries from physical injury while providing good ventilation for rapid cooling.

Trays are grouped together in a two-high stack with vertical wires at both ends of the trays to keep the pallet stable. The wires protrude above the top tray and fit into slots in the bottom of the tray stacked above. Fiberboard tie sheets are placed on top of the sixth, tenth, and top tray layers. Wires and tie sheets stabilize the pallet load and protect the fruit such that pallet wraps and strapping typically are not necessary.

Berries are picked with their caps (calyxes) attached. Some special "stem-grade" fruit are picked

Figure 11.1. A strawberry harvest operation. By having each picker pick from only half the bed to each side, you eliminate excessive reaching and simplify the selection, sorting, and placement of the berries into trays on the wheeled cart.

Figure 11.2. Package design holding eight 0.45-kilogram (1 lb) clamshell baskets made of clear, thermoformed plastic. *Photo* by J. Thompson.

Figure 11.3. Common package design holding 4 to 5 kilograms (9 to 11 lb) of fruit in twelve 1-dry-pint polypropylene mesh baskets. *Photo* by J. Thompson.

by pinching the stem about 5 centimeters (2 inches) from the calyx. The fruit must be held loosely in the hand without squeezing, since any squeezing of the fruit will cause bruising injury and discoloration. Strawberries should be placed (not dropped) into the crate. While the crates need to be well filled for marketing, they should not be so full that berries in one crate will be crushed when another crate is stacked on top. Berries can be cut by the rim of open-mesh baskets as a result of being packed over the top of the baskets. Injuries from this type of packing cause both a direct loss from physical damage and an increased risk of Botrytis rot (gray mold), which can spread to undamaged berries. The clamshells eliminate this type of damage.

PREPARING HARVESTED STRAWBERRIES FOR MARKET

Organic strawberries and conventionally grown strawberries are handled similarly after harvest. With organic fruit, however, fewer options for disease control during fruit production combine with the lack of any residual protection from conventional preharvest fungicide applications to increase the importance of postharvest temperature management and other decay control strategies.

Temperature Management

Good temperature management, including rapid cooling and the maintenance of low fruit temperatures, is the single most important factor in protecting harvested strawberries, to help minimize decay and overripening and to maximize their postharvest life. Temperature conditions and the length of exposure to those conditions together determine the amount of deterioration that is likely to occur. Because of this relationship, a constant circulation of delivery trucks to and from the harvest field is required, delivering frequent, small loads to the cooler. Arriving fruit should be moved promptly to the cooler, without delays for inspection or other processing. A shaded receiving area is recommended to prevent sun exposure that can warm fruit to temperatures well above air temperature during fruit unloading.

Because of the harmful effects of cooling delays, it is recommended that cooling begin within approximately 1 hour of picking to prevent decreases in the percentage of marketable fruit. Forced-air cooling is highly recommended for removing field heat and cooling the fruit to storage temperature as quickly as possible. Forced-air cooling is a specific method of cold air management that positions pallets of fruit so that cold air must pass through package openings and around individual berries (figure 11.4).

The design of the strawberry crates commonly used in California and Florida, with their large side ventilation openings, allows a modest air pressure difference to move large volumes of air across packed berries, making berry cooling quite rapid (Talbot, Brecht, and Sargent 1995). Approximate airflow characteristics needed to cool strawberries in standard corrugated crates with mesh baskets are shown in table 11.1. Cooling rates for mesh baskets and pint-sized clamshells are similar, but quart clamshells take slightly longer to cool. The amount of venting on the clamshell baskets should be close to 13 percent of the side surfaces.

"Seven-eighths cooling time" is the time it takes to cool the berries by seven-eighths of the difference between their initial temperature and the temperature of the cold air. A 24°C (75°F) strawberry in –1°C (30°F) air is seven-eighths cool when it reaches 2°C (about 36°F). All of these values are based on the warmest fruit in the pallet (the fruit on the downstream position inside the cooling tunnel). The pallet farthest from the fan will usually contain the warmest fruit.

Where no forced-air cooling is available, trays of fruit should be spaced out in the cold room to increase their rate of cooling. Use circulation fans in the cold room to increase air movement and speed cooling.

Table 11.1. Airflow characteristics and time to cool strawberries using forced-air cooling

	Airflow (cfm/lb berries)			
	2.0	1.4	0.8	0.5
Static air pressure (psi)	0.40	0.20	0.08	0.04
Hours to ⅞-cool warmest fruit	1.5	2.0	3.0	4.0

Figure 11.4. Forced-air cooling of strawberry pallets. Pallets are placed in two rows on alternate sides of an exhaust fan. A tarp is placed over the top to make a tunnel, and the cold air is pulled through the sides of the strawberry crates, across the fruit, and to the exhaust fan.

Refrigerated Holding

Depending upon their handling and their harvest maturity, strawberries may have a market life of from 1 to 2 weeks. Strawberry coolers and holding facilities should be maintained as close to 0°C (32°F) as possible, with minimum temperature fluctuation. The highest freezing point that has been reported for strawberries is –0.8°C (30.6°F), but this is dependent upon their soluble solids concentration: fruit with higher soluble solids have a lower freezing point. By maintaining high relative humidity in storage areas, you can minimize water loss. An unsealed pallet cover can maintain higher humidity around the fruit, preventing dehydration and maintaining the fruit's glossy appearance.

Carbon Dioxide Treatment

Many strawberries are shipped with an elevated–carbon dioxide treatment, sometimes called modified-atmosphere transport (Kader 1991). This may help slow the physiological activity of the berries, thus slowing their rate of deterioration. The main purpose of modified-atmosphere transport is to reduce the spread and development of rots, especially gray mold. This preventive effect on gray mold development is measurable at transport temperatures above about 2°C (36°F), when growth and spread of the rot are more active. The value of carbon dioxide treatment will depend on the transport time and temperature and the decay potential of the fruit. The greatest benefit would be expected for berries harvested after periods of cool, moist or foggy weather when free water may have collected on berries in the field and the spread of gray mold would be expected to be most severe.

The standard method for carbon dioxide treatment is to completely enclose a pallet load of cold berries in a sealed plastic bag, create a slight vacuum in the bag, and then add carbon dioxide to create a 12 to 15 percent carbon dioxide atmosphere within the pallet bag and around the fruit (figure 11.5). It is important that the fruit be thoroughly cooled before bagging since the plastic pallet cover impedes further cooling, and condensation will form inside the bag if the berries are not fully cooled. All preparation and treatment should occur inside the refrigerated holding room, and pallet covers should be applied just prior to truck loading. The temperature of the berries shipped under a pallet cover may be 2°C (4°F) warmer than for uncovered berries, so you need to take special care to protect the bag from tearing, as that would result in the loss of the protective carbon dioxide atmosphere. If the bag does tear, strawberry quality can deteriorate more quickly within the bag due to the slightly warmer temperature.

Other strawberry pallet covers on the market are not sealed at the bottom and do not modify the atmosphere within the pallet. The main benefit of these covers is to reduce water loss by maintaining a higher relative humidity.

Figure 11.5. (A) A pallet bag is placed over a pallet of thoroughly cooled strawberries. The bag will be sealed to the plastic pallet liner under the fruit. (B) With the plastic bag and pallet liner carefully taped together, a nozzle pierces the bag to exhaust some air and then replace it with carbon dioxide (CO^2). Tape will cover the point of injection in order to maintain the gas seal.

TRANSPORTING FRESH STRAWBERRIES TO MARKET

The period when fresh strawberries are in transport makes up a major portion of their total postharvest life. This means that good loading and transport conditions, including maintenance of low fruit temperatures, are critical to the successful marketing of fresh strawberries.

Refrigerated Loading

Refrigerated holding rooms or loading docks for strawberries should be equipped with sealed loading doors that allow berries to remain under refrigeration while they are loaded into transport vehicles. The transport vehicle should be thoroughly inspected to ensure that the refrigeration and airflow systems are operating properly and that the vehicle and any contents have been completely cooled before strawberry loading begins.

The location of the load within a trailer is not a concern so long as the trailer is equipped with air-ride suspension, as is more commonly seen today. If the rear trailer axles have normal spring suspension, however, damage to strawberries during transport is less likely if the fruit are placed near the front and center of the vehicle, since that puts the tractor axle under the load and that axle is almost always equipped with air-ride suspension for driver comfort.

Past studies have shown that considerable heat can penetrate trailer walls and reach fruit when strawberries are loaded against side walls during cross-country transport, especially during warm summer months. To prevent this, load the pallets in the center of the trailer, with dunnage blocks along the sides between walls and pallets to stabilize the load. This ensures that there will be space for airflow between the trailer walls and the pallets.

Setting the Thermostat

The vehicle thermostat should be set to as low a temperature as is possible without risk of causing freezing damage to the strawberries. If the fruit's freezing point is near –0.8°C (30.6°F) (this depends on the soluble solids content of the fruit) and if the accuracy of the vehicle's thermostatic control equipment is ± 1.5°C (± 3°F), then you may be able to set the thermostat as low as 1°C (34°F). Any other produce to be loaded with the strawberries should be compatible with the same temperature requirements.

Air Transport

Strawberries are often transported by air, particularly to export markets. Strawberries for air transport should receive the same preparation and protection as strawberries for surface transport. Berries should be cooled and loaded as discussed above, and the fruit should be kept cold for as long as possible before being staged for air cargo handling. Efforts should be made to minimize exposure of strawberries (in pallets or air-freight containers) to open sunlight, especially on hot tarmac staging areas. Reflective pallet covers with cold packs may help reduce warming. Tests have shown that even though strawberries may warm during transport, it is better to provide cooling whenever possible than not to cool them at all. Upon arrival at their air destination, pallet covers should be removed and the fruit should be returned promptly to refrigeration, using forced air if possible to re-cool them to near 0°C (32°F) as quickly as possible.

CARE OF STRAWBERRIES IN DISTRIBUTION AND RETAILING

Receiving

Cold fruit in refrigerated vehicles should be unloaded directly into the 0°C (32°F) warehouse whenever possible. If the fruit are warmer than about 2°C (36°F), they will benefit from being re-cooled and held at 0°C (32°F). If inspection indicates the presence of decay and the trays need to be sorted and repacked, the fruit should still be placed at cold temperatures and small lots removed momentarily for sorting.

Handling Bagged Pallets

Strawberries in pallets that have been bagged (whether for carbon dioxide treatment or to reduce water loss) will warm somewhat during transport as a result of their own respiration. The bags do not allow cold air to flow through the pallets to remove this respiratory heat. It is likely that some of the modified-atmosphere pallet covers will have been punctured during loading and unloading and that their high–carbon dioxide atmosphere will have been lost, so it is important that both types of pallet covers be removed and the berries cooled thoroughly upon receipt.

Refrigerated Holding

Strawberries in a distribution facility should be held between 0° and 2°C (32° and 36°F) under reasonably high (90 to 95%) relative humidity. In low–relative humidity air, the fruit would readily lose water and quickly shrivel and appear older than they are. Loss of vitamin C has also been observed in strawberries stored under low–relative humidity air. Because shriveling of fruit is a result of cumulative water loss, fruit that appear sound may still be approaching the stage at which shrivel becomes visible when they arrive at the distribution market.

Loading for Retail Store Distribution

Because of their highly perishable nature, strawberries should receive special handling in their final stage of distribution. They should not be allowed to accumulate in warm staging areas where individual store orders are assembled, nor should they be loaded into warm distribution vehicles awaiting the assembly of other products to complete the load. Whenever possible, strawberries should be kept cold until the delivery truck is ready to depart, and only then placed into the load for final movement to retail outlets.

CONCLUSIONS

Organic strawberries are handled very much like conventionally grown strawberries after harvest, but they may have a greater potential for postharvest decay due to limitations on the disease management options available to growers during fruit production and the lack of residual protection from preharvest fungicide applications. For this reason, it is especially important that organic growers and shippers take extra care in postharvest handling. Gentle handling, rapid and prompt cooling to 0°C (32°F), and shipment under elevated–carbon dioxide atmospheres are the critical strategies for maintaining fruit quality after harvest. And, of course, the development of additional tools for control of postharvest decay in organic strawberries would be beneficial.

REFERENCES/RESOURCES

Kader, A. A. 1991. Quality and its maintenance in relation to the postharvest physiology of strawberry. Pp. 145–152 in A. Dale and J. J. Luby (eds.), The strawberry in the 21st century. Portland, Oregon: Timber Press.

Mitchell, F. G., E. Mitcham, J. E. Thompson, and N. Welch. 1996. Handling strawberries for fresh market. Oakland: University of California Division of Agriculture and Natural Resources. Publication 2442.

Talbot, M. T., J. K. Brecht, and S. A. Sargent. 1995. Cooling performance evaluation of strawberry containers. Proceedings of the Florida State Horticultural Society 108:258–268.

CERTIFICATION ISSUES

PHOTO: ISTOCKPHOTO.COM

CERTIFICATION ISSUES

12 Organic Certification and Registration in California

DAVID CHANEY, RAY GREEN, L. ANN THRUPP, VANESSA BOGENHOLM, AND SEAN L. SWEZEY

The organic industry has grown significantly during the 1990s and early 2000s. For the United States as a whole, the total market value (retail sales) of organic food products, including processed products, grew from about $1 billion in 1990 to an estimated $21.1 billion in 2008 (Dimitri and Greene 2002; USDA ERS 2011), while the Organic Trade Association estimated 2009 total organic product sales in the United States at $26.6 billion (OTA 2010). California producers have led this trend, showing an increase in both numbers of organic farmers and total acreage. From 1992 to 2008, the number of registered organic farms in California more than doubled, from 1,273 to 2,961 (Klonsky 2010). Over the same period, organic acreage grew tenfold, increasing from 42,000 acres in 1992 to 470,700 acres in 2008 (Klonsky 2010; Klonsky and Tourte 1998; Klonsky et al. 2002; Klonsky and Richter 2005).

Prior to 1990, few common labeling standards were agreed upon, and for the most part there was no regulatory requirement for organic certification. As the market began expanding, organic producers and marketers recognized the need for more uniformity and integrity in their products, and they turned to Congress for assistance in developing national standards to govern organics. The result of these efforts was the Organic Foods Production Act (OFPA) passed in 1990 as part of the Food, Agriculture, Conservation, and Trade Act. OFPA mandated that the U.S. Department of Agriculture (USDA) establish an organic certification program for producers and handlers of agricultural products who sell and label their products as being organic. Responsibility for developing and administering a National Organic Program (NOP) was assigned to USDA's Agricultural Marketing Service (AMS). The intentions of OFPA were to establish national practice standards governing the production and marketing of certain organically produced agricultural products, to assure consumers that organically labeled foods represented these consistent standards, and to facilitate interstate commerce in fresh and processed food that is organically produced and labeled as such.

California did have an organic labeling law in place starting in the 1970s, but the law had no enforcement component. Concurrent with the development of OFPA, California passed its own law regulating organic farming, the California Organic Food Act (COFA), which was signed into law in 1990. COFA set a voluntary certification standard with full registration and enforcement components. The California law became effective immediately upon signing, but it took more than 10 years for the federal OFPA to be completed and implemented. When COFA was eventually superseded by the federal regulations, it was rewritten to become the California Organic Products Act of 2003 (COPA).

The process of completing the national rule was complicated and controversial, involving numerous drafts and public input from consumers and the organic industry. USDA issued a proposed rule in 1997, but the final program was not completed until December 2000. The final rule states that all organic producers who gross more than $5,000 per year in retail sales must be certified through an accredited certification agency. These certifying organizations, both public and private, act as third-party agents of the NOP to legally verify that production and handling practices meet the national standards. Enforcement of the NOP began in October 2002, and AMS immediately began to accredit certification agencies to assist with its implementation.

In addition to accreditation of certifiers, the NOP is also responsible for ensuring that the purposes of the OFPA are accomplished, determining the equivalency of foreign programs for imports into the

United States, participating in development of international standards, coordinating enforcement activities with other agencies, conducting the petition process for materials review, and providing administrative support for the National Organic Standards Board (NOSB). The NOSB is a 15-member committee established as part of OFPA to help the Secretary of Agriculture develop standards for substances to be used in organic production and to advise the Secretary on other aspects of implementing the program. The current board includes four farmers/growers, two handlers/processors, one retailer, one scientist, three consumer/public interest advocates, three environmentalists, and one certifying agent.

STATES AND THE NOP

State governments can serve two functions in relation to the NOP: they can choose to implement a State Organic Program (SOP) or to become an accredited certifying agent, or they can choose to do both. An SOP is primarily an enforcement program and is not involved in certifying growers or handlers. Once approved as an SOP, the agency is responsible for all aspects of enforcement within the state. A state agency that becomes an accredited certifying agent serves the same function—certifying organic production, processing and handling—as any other certifying agency would. In the list below, you can see how three states are working in this area.

- *California.* The California Department of Food and Agriculture (CDFA) administers an SOP only and does not certify.
- *Washington.* The Washington State Department of Agriculture is an accredited certification agency only, not an SOP.
- *Utah.* The Utah Department of Agriculture and Food administers an SOP and is also an accredited certifying agent.

In order to qualify as an SOP, a state must adopt its own set of regulations or laws that accept the standards set by the NOP. Upon approval by USDA, those state laws essentially become the NOP laws for that state. COPA (the current California law), for example, incorporates by reference the federal regulations and is interpreted in conjunction with them, and it is written so that any future changes to the federal law automatically become regulations for California. Compliance and enforcement provisions are part of both the NOP and approved SOPs. Most states do not have SOPs; they fall under the direct jurisdiction of the federal NOP with regard to enforcement activities.

NOP IMPLEMENTATION IN CALIFORNIA

California's SOP is administered through CDFA (responsible for fresh agricultural commodities [e.g., fruits, nuts, vegetables, field crops], meat, poultry, and dairy products) and the California Department of Public Health (CDPH), which is responsible for processed food products, cosmetics, and pet food. The SOP's purpose is to protect producers, handlers, processors, retailers, and consumers of organic foods sold in California by enforcing labeling laws related to organic claims for agricultural products. Its activities are coordinated with the California Organic Products Advisory Committee, USDA, and the California county agricultural commissioners.

The California SOP's responsibilities include administering the program, training county staff, investigating complaints, registering producers and certification organizations, and acting as a resource for information on COPA and the organic industry in California. The SOP is funded entirely by producer and certifier registration fees, a portion of which are used to support county enforcement activities. Because COPA brings two separate statutes relating to fresh and processed products together into one law, it is important that each individual carefully read the sections that relate to his or her type of business operation. The complete text of COPA is available online at the CDFA Organic Program website, www.cdfa.ca.gov/is/i_&_c/organic.html. If you are involved in both aspects—growing and processing—you should note the specific requirements pertaining to each.

Producers

Organic producers and handlers (e.g., wholesale distributors, retailers, and some processors who market direct to retail or to consumers) are regulated through CDFA. California requires that organic producers be registered and, depending on their sales revenue, certified as well. All organic growers, regardless of gross sales, must be registered with CDFA. In addition, any producer who grosses more than $5,000 per year in total sales of organic products must be certified through an accredited certification agency.

Registration, which involves providing a map of the production area, a list of crops that you intend to produce as organic, and a 3-year history of substances or materials applied, is handled by your local agricultural commissioner. You have to provide verification of the land use history and pay an initial minimum registration fee of $75 (the fee is higher if annual gross sales are—or are expected to be—in excess of $25,000; see table 12.1). After initial registration, fees for yearly renewal follow the schedule shown in table 12.1.

Certification follows a separate process. NOP regulations establish who must become certified before they can use the "organic" label on their products. The following general guidelines apply:

- *Exempt from having to be certified.* Producers who are grossing less than $5,000 per year in retail sales are exempt from having to be certified. Product must be produced, processed, and packaged under the producer's own label; product cannot be sold to anyone else for processing or packaging, although the product can be marketed in retail stores under the producer's own label.
- *Excluded from requirement of having to be certified.* Any operation or company that does not repackage or re-label the product (e.g., bulk wholesale distributor) or retail establishments that process or prepare organic product on the same premises at which they sell to the customer is excluded from the requirement of having to be certified.
- *Must be certified.* All other operations involved in production, processing, and marketing must be certified.

If you are not sure whether you need to be certified or what category you fall into, contact the CDFA SOP office at (916) 445-2180 (online at www.cdfa.ca.gov/is/i_&_c/organic.html). The steps involved in certification are described in more detail under "Certification Agents," later in this chapter.

Processors

As defined in COPA, every person in California who processes, packages, stores, distributes, or handles food, pet food, or cosmetics in California that are sold as organic (with the exception of processed meat, fowl, or dairy products) is required to register with the California Department of Public Health (CDPH). Registration is handled through the CDPH's Organic Processed Product Registration Program, and processors must pay an annual registration fee. Like producers, they must be both registered and certified.

CDPH itself handles the registration of organic processors. To get an application, call CDPH or visit the CDPH website (www.cdph.ca.gov/programs/pages/FDB.aspx). The application must be filled out completely and returned to the CDPH with the appropriate fee. Upon receipt of the application and fee, the operation is registered, provided that if the operation processes food for human consumption, the operation already has a valid processed food registration permit issued by CDPH. There is no preinspection or verification requirement, but during the operation's regular food safety inspection the inspector may ask questions about the organic aspects of the operation. The registration fee is set according to the schedule shown in table 12.2. Fees for yearly renewal follow the same schedule.

Table 12.1. Fee schedule for companies required to be registered with CDFA

Gross sales	Annual registration fee
$0–$4,999	$25
$5,000–$10,000	$50
$10,001–$25,000	$75
$25,001–$50,000	$100
$50,001–$100,000	$175
$100,001–$250,000	$300
$250,001–$500,000	$450
$500,001–$1,000,000	$750
$1,000,001–$2,500,000	$1,000
$2,500,001–$5,000,000	$1,500
$5,000,001–$15,000,000	$2,000
$15,000,001–$25,000,000	$2,500
$ 25,000,001 and above	$3,000

Note: Fees are subject to change; check with CDFA for the most current fee schedule. Producers who sell processed product pay fees based on the value of raw product prior to processing and the value of any product sold as unprocessed. Other exceptions are outlined in the text of COPA (see the CDFA Organic Program website, www.cdfa.ca.gov/is/i_&_c/organic.html).

Table 12.2. Fee schedule for companies required to be registered with the California Department of Public Health

Gross annual sales or revenue	Annual registration fee
$0-$5,000	$50
$5,001-$50,000	$100
$50,001-$125,000	$200
$125,001-$250,000	$300
$250,001-$500,000	$400
$500,001-$1,500,000	$500
$1,500,001-$2,500,000	$600
$2,500,001 and above	$700

Note: Fees are subject to change; check with the CDPH or CDFA for current fee schedule.

It is possible that an operation or business will have to register both as producer and processor with both agencies (CDFA and CDPH). For example, strawberry growers who also make and sell a processed product, such as strawberry jam, must be registered with both CDFA and CDPH. The registration fee for CDFA is based on the market value of the strawberries prior to processing, whereas the fee for CDPH is based on the market value (gross sales) of the strawberry jam.

Certification, which is distinct from registration, involves the same steps and requirements for processors as it does for growers (see "Certification Agents," this page). The same restrictions and exemptions described above for producers also apply to processors. If your farming business is involved in both production and processing, it makes sense for you to find a certifying agency that is accredited to certify both aspects (production and processing) of the operation. Most certifiers do both, but it is wise to make sure. Some certifiers consider postharvest handling to be a separate activity requiring a separate application, fee, and inspections. Other certifiers see processing as an addendum or attachment to the production side of the operation, an approach that can reduce costs and paperwork for the grower. This discretion on the part of the certifier only goes so far: any activities beyond simple postharvest handling (e.g., beyond storage or transportation to market) would clearly fall under the "processing" rules and require separate registration and certification for each component.

If you are not sure whether you need to be certified or what category applies to your operation, contact the CDPH Food Safety Program at (916) 650-6500. The remainder of this chapter covers the certification process and is written mainly for growers, but the general principles and steps involved in becoming certified apply to processors as well.

Certification Agents

An organic producer can choose from any of the certification agencies accredited by USDA and should select the one that will best serve his or her needs and budget (see sidebar, "How to Evaluate and Select a Certifier"). As of spring 2008, more than 90 accredited certification agents were approved by the USDA, including state and international agencies as well as private companies. About 20 of those are registered under the California SOP. All certifiers who want to operate in California must register with CDFA. Most certifiers require an application fee as well as inspection fees. These are substantial, starting at around $500. If you intend to export your operation's organic products outside of the United States, you will also need certification from the certifying body in the destination country or from the International Federation of Organic Agricultural Movements (IFOAM). Some certifiers that are accredited through the NOP have direct partnerships with those foreign certifiers, meaning that they are accredited by those international agents and can verify international certification in addition to NOP certification.

The certifying agency monitors grower practices to ensure that the grower is in compliance with NOP regulations. The certifier

- MUST know and enforce the NOP standards.
- MUST verify the client's compliance with national standards through annual scheduled inspections of his or her records, fields, and production and handling areas.
- MAY perform annual soil and tissue tests from the client's operation in accordance with NOP guidelines.
- MAY perform a certain number of surprise inspections every year.
- MUST conduct ongoing review and inspection of client operations to ensure compliance.

As a third-party agent of the NOP, the certifier

HOW TO EVALUATE AND SELECT A CERTIFIER

Questions to ask yourself before you begin contacting certifying agents:

- *Will my product be sold only within California?*
  If so, you may want to contact only agents with home offices in California. If the product will be sold outside California, other certifiers including those based in other states can be considered.
- *Will my product be sold outside the United States?*
  If not, you need only NOP certification. If you are exporting to other countries, then you should look for agents who certify according to international standards, such as the Japanese Agricultural Standard (JAS), the European Union (EU), or the International Federation of Organic Agriculture Movements (IFOAM), and who have knowledge of those markets.
- *Is my product a raw crop or a processed food?*
  Which agents seem to be strongest and which seem to provide the best service in relation to my final product?
  Do some research on the certifiers you are considering.
- *Who certifies my friends and other companies like mine in the county or state?*
  Talk to those companies and find out how satisfied they are with their certifying agents.

Questions to ask certifying agents after you have checked the NOP website and confirmed that the certifying agent is accredited by the NOP:

- *Given the products, land, or processing that I propose to certify, what are the expected costs for the initial submission of a certification request, actual certification on the first inspection, and expected ongoing annual costs? When are the fees due and payable?*
- *How long have you been certifying organic operations?*
- *Do you offer training programs in any aspect of organic compliance?*
- *How do you determine which materials are allowed and which are prohibited?*
- *Is your cost or fee structure linked to organic sales, overall sales, number of acres, or complexity of the operation?*
- *Once paid, are any of my fees refundable if I decide to withdraw from application or from certification? Are there any additional costs or hidden fees?*
- *Are there any discounts for renewal clients?*
- *How many pages would a typical application package be for an operation similar to mine?*
- *I am going to export to a foreign country (give the name of the country or market). What is your expertise in certifying for this target market? What documents or transaction forms are required for export of organic product to this country? Are there any fees for generation of specific documentation required for each individual shipment?*
- *What are your policies regarding confidentiality of information and records?*
- *What are the charges if I need to add or change something (land, crop, product, SKU of same product, etc.) during the year, and what documentation is required?*
- *Will you assist in any manner in the marketing of organic commodities?*
- *Do you have a website listing the companies you have certified?*
  Do you list companies you have certified by commodity?
- *Do you certify co-handlers under my application? In that situation, is the fee reduced?*
- *What type of verification is needed to prove land history?*

Other considerations:

Every certifier is a private company or is part of a government entity. Each has a different ability to respond to you and to provide services. Check the level of service by finding out the following:

- *What is the average time that it takes the certification agency to complete a certification?*
- *What is the average time that it takes the agent to call back or respond to a question you may have?*
- *Do they provide supporting information that helps you understand compliance issues?*
- *What kind of support does the certifier offer to help you understand and complete paperwork?*
- *Does the certifier provide a detailed explanation outlining the sections of the NOP that back up their certification positions and decisions?*
- *Are the inspectors friendly and courteous? Are the inspectors actual employees of the certifying agency or does the agency contract with independent inspectors? Do you have the option of choosing your inspector?*

Notes:

The NOP's official list of accredited certification agencies is published on its website, www.ams.usda.gov/AMSv1.0/nop. *The New Farm's Guide to U.S. Certifiers* is a very useful tool for learning about and comparing accredited certification agencies in the United States. For more information, visit the *New Farm's Guide* website, http://newfarm.rodaleinstitute.org/ocdbt/.

is forbidden to provide advice on the producer's operation, but can and must provide information to the producer about the certification process and the legal requirements for maintaining certification. This means that the certifier should know the client's operation in detail and should have current knowledge of NOP regulations and be able to communicate them effectively, but may not act as a consultant to the grower. Certification agents must maintain their accreditation through the NOP and, as noted in the sidebar, may also become certification agents for foreign or international bodies. Some certifiers may also serve a broader role in organic trade associations or in dealing with policy issues.

THE ORGANIC SYSTEM PLAN AND GROWERS' RESPONSIBILITIES

Growers should recognize that they also have serious responsibilities and commitments in relation to certification. According to NOP regulations, every certified organic producer is required to develop and keep current a production or handling organic system plan (OSP) that describes the operation and the practices and procedures to be maintained and also provides a list of each substance to be used as a production or handling input. The OSP includes

- land history
- sources of all seed and planting stocks used by the organic operation
- all inputs to the crop and all materials applied to the crop
- a list of all practices and procedures for soil and pest management, fertility, and crop nutrition that are used in the operation, including details on monitoring practices
- an explanation of barriers or buffers used to prevent commingling between organic and nonorganic products or drift from nonorganic operations
- harvest and postharvest practices, including equipment use
- a description of records kept to prove compliance
- any additional information needed to document NOP compliance

Most certifiers provide a form that growers can use to prepare their OSP. These forms may be helpful for some growers, but they are not required; it is possible to write your own OSP, so long as you include all of the required information. In either case, writing the OSP is useful for planning your farming operations for the year. Some certifiers may provide help in preparation of the OSP. The length of time to complete an OSP depends on the level of complexity of the farming operation.

The OSP is central to the certification process. It serves as a management tool to help farmers make decisions and react to changing circumstances. It also describes the human and natural resources of a farm, helps a producer manage those resources in an integrated way, and can help the grower with budgeting and financial planning. Last, and most important, the OSP constitutes a legally binding contract between the certifier and the certified operation. Breach of that contract can result in denial or loss of certification. This last point is crucial, and growers must understand that the records they keep constitute the only proof that their "contract" (the OSP) has been fulfilled. The OSP is the commitment or promise to the certifier that production and handling will be carried out in a certain way, but the grower must show through good record keeping how he or she has kept that promise. Growers must get approval from their certifier if they are going to deviate from their submitted organic system plan.

In summary, the federal OFPA is a performance regulation, not a prescriptive regulation. This means producers decide how they will grow or process their products and how they will demonstrate compliance with the NOP. The OSP is the individual grower's or operator's specific plan on how his or her farm is going to meet those performance targets. Growers should be proactive and create the roadmap for achieving compliance that works best for them.

TRANSITION AND STEPS FOR ORGANIC CERTIFICATION

Transition Period

In order to be certified organic, a farming site cannot have had any prohibited substances applied to it for 36 months. One way to accomplish this is to choose a site that has not been farmed for at least the last 3 years. If the site has been actively farmed, the grower must ensure (and verify with accurate records) that no synthetic, noncompliant materials were used during the 3-year period. The 3-year transition period is a time when growers can educate themselves about the NOP and certification requirements (see the list of Internet resources at the end of this chapter). This is also an opportunity to begin work on a soil-building program, an integral component of the OSP. A grower in transition may also consider hiring a consultant to help prepare the farm for certification. This is not required and adds to the transition's cost, but a good consultant can help the grower avoid mistakes that might otherwise slow the farm's progress toward certification.

Upon obtaining certification, the grower must implement all production practices addressed in the regulation and must be proactive in building up soil organic matter with compost, cover crops, or other amendments, using organically acceptable methods to control weeds, insects, diseases, and any other pests, and implementing biodiversity. The grower must document in detail the name and amount of every material applied to the field and must keep the OSP current. A producer does not need to be certified or registered during the transition period.

Keeping Records from the Start

It is very important that growers keep careful records of the exact date they began the organic transition, the last prohibited materials applied to the field and when they were applied, and all of the specific inputs and practices used during the 3-year transition period. When you apply for certification, the certifier will need detailed production practice records in all those areas. We recommend that you keep receipts and labels of organic inputs and that you get verification from neighbors, PCAs, county agricultural commissioners' pesticide use reports, or information from other local officials to corroborate that you ceased using prohibited pesticides (and when you did it) and when you started the organic transition. Also, it is helpful to keep records of production activities or tasks by recording work orders or maintaining other labor records.

Contacting the Certifier

The certifier will send you all of the application materials you will need to get started. Certifiers expect growers to read the handbooks or literature they provide. The grower's OSP must be fully prepared and submitted with the appropriate forms. Many certifiers require that growers sign an affidavit that shows land use history and affirms the truthfulness of the application. It usually takes 2 to 5 months for the application to be reviewed. Certifiers charge certification fees, which generally include

- a one-time application fee (ranging from $100 to $300)
- inspection costs, which usually pay for an inspector's time to travel to and inspect the farm and prepare a report, usually once per year
- annual fees (some certifiers use a fee based on acreage; others charge an annual fee based on some percentage of the gross production value)

Make sure you know the specific fees that the certifier will charge before you apply. The application fee must be paid when you submit the application. The fees cover the costs of the certification procedures and agency overhead and pay for the time of the inspection staff. Through certification, growers gain transparent third-party verification of their compliance with the NOP. This does open the operation to greater public scrutiny, but ultimately it enhances the value of the farm's product in the marketplace.

Review of Application, Inspection, and Notification of Certification

The certifier will review the grower's application and inspect the operation before certification is approved. First, a trained inspector calls the grower to set up the initial inspection. After the inspection, the inspector submits a report to the certifier for review. The certifier informs the grower of the certification status or lets the grower know about any further requirements the grower must meet in order to achieve or maintain certification.

During an inspection, the inspector will want to review or see records about

- land use history
- pesticide use reports
- cleaning or cleanout procedures if you use shared equipment
- your understanding of the regulations (i.e., verification that you have read the appropriate manuals)

- the use of approved materials only
- appropriate systems to ensure compliance
- product labels for products or inputs that you use

Annual Review and Inspection

Organic certification inspections are done on farms once a year as required by the NOP. The inspector generally informs the grower of each visit in advance and sets up an appointment for the inspection. Certifiers and state organic inspectors do, however, have the right to make unannounced visits or inspections as well. If an inspector encounters violations or problems, the grower will receive a notice of minor noncompliance or major noncompliance from the certifier. If the noncompliance is minor, the grower is reprimanded, told not to repeat the error, and may be sent a reminder about the particular issue. If the inspector considers the noncompliance to be major, growers' certification status may be affected. The grower may be required to submit requested documents or update a section of the OSP, or in serious cases the grower can be decertified.

As discussed previously, certification inspectors are not allowed to provide advice or information to the growers while conducting an inspection. They do not serve as consultants or advisors. This is intended to ensure that the certifier's role is enforcement and compliance checking. However, some certification agencies do have departments or divisions that provide information or educational materials to all growers, separate from the certification enforcement. In addition, many organizations, services, advisors, websites, and other resources provide information about organic practices.

A summary of the certification process and the interaction between the grower and the certifier is outlined in table 12.3.

ISSUES REGARDING COMPLIANT AND NONCOMPLIANT MATERIALS

All materials that can be used in crop and livestock production are classified into one of four categories under the NOP. Note that in the context of the NOP regulation, the words "nonsynthetic" and "natural" are used synonymously.

- *Category 1: Allowed nonsynthetic.* Most nonsynthetic (natural) materials are allowed, with the exception of those that are specifically prohibited (category 2, below). There is no list of allowed natural materials; the rule of thumb is that natural materials are allowed unless they are specifically prohibited.
- *Category 2: Prohibited nonsynthetic.* A few nonsynthetic (natural) materials, such as arsenic and strychnine, are not allowed under the NOP. See NOP Rule §§ 205.602 and 205.604.
- *Category 3: Allowed synthetic.* These synthetics are allowed, but their use is restricted to specific purposes as defined in the NOP Regulation and National List. Use of these materials for other purposes is prohibited. See NOP Rule §§ 205.601 and 205.603.
- *Category 4: Prohibited synthetic.* Most synthetic materials are not allowed. The exceptions are those specifically approved for use in organic production (category 3, above).

The National List of allowed materials for organic production is overseen and maintained by the NOP (www.ams.usda.gov/AMSv1.0/nop). The National List indicates generic compounds and materials that are allowed but does not list specific brand name products. Brand name information is available through the Washington State Department of Agriculture (http://agr.wa.gov/Portals/Org/) and the Organic Materials Review Institute (www.omri.org). OMRI is a private, nonprofit organization that provides verification and listings of products that meet the national organic standards. They publish and regularly update two lists: a brand name list and a generic list.

As a certified grower, you are expected to keep track of updates and changes in the approval status of the materials you use. Before trying new materials, you should research the NOP, OMRI, and WSDA lists to make sure that the materials are allowed and also get permission from your certifier to amend your OSP and use the product. Do not rely solely on the verbal or written declarations of vendors. Consult a reliable source to confirm any vendor claims.

The certification agency has the final decision on the acceptability of inputs for each farm operation. In making a determination about the acceptability of an input, the certification agency must evaluate the input based both on its ingredients and on the

Table 12.3. Roles and responsibilities of growers and certifiers

Grower	Certifier
During transition	
Read and understand NOP regulations. Gather information. Do research and planning. Begin transition.	
Document NOP compliant practices and management for 3 years. Select accredited certification agent.	Certifier would have a role only if grower seeks transitional certification.
Request application materials and instructions from certifier.	Provide materials and information to client (e.g., NOP regulations, application materials, etc.).
After 3 years	
Register with CDFA prior to sale of organic products.	Certifier not involved in registration—grower registers through CDFA.
Submit application and organic system plan for certification to certifier (recommend this be done well in advance of first anticipated organic harvest [6 months to 1 year ahead]).	Verify that grower is registered with CDFA and review application for certification in a timely manner. Request additional information as necessary from grower.
Supply any new information or details requested by certifier to demonstrate compliance. Grower must submit to on-site inspection.	Conduct inspection and notify grower of results. Upon approval, OSP becomes a legally binding contract.
Implement OSP. Record practices and procedures. Obtain prior approval for any alterations to the OSP. Keep accurate and detailed records. Keep records organized in preparation for annual inspection.	Annually review and inspect grower's application for assurance of certification compliance.

context in which it will be applied. Prior to using any input on an organic farm, you must obtain written approval (e.g., a certificate of compliance) from your certifier. The certifier will contact you or the material's manufacturer if he or she needs any additional information to determine whether use of the material is in compliance. A word of caution: any input that you use in your farming operation without the written approval of the certifier could be viewed as a departure from your agreed-upon OSP and could be grounds for adverse action on the part of the certifier, USDA, or state programs.

CERTIFICATION ISSUES FOR STRAWBERRY GROWERS

When initiating a management and business plan for the production of certified organic strawberries, it is essential that you know what information will be required of you by the certifying agent for inspection. Initially, the organic strawberry grower must decide on the intended market and handling details for the organic strawberries. Are you planning to produce organic strawberries for domestic sales only, or are you considering export to Europe or Japan? If foreign destination export is part of your plan, you will have to meet the certification standards of the receiving country. In many cases, these will be different, and possibly more detailed, than U.S. standards. A certifier familiar with your export markets will have to prepare an inspection report that indicates compliance with the standards of the destination country.

If you are using an outside sales company or cooler, these companies and facilities also have to be registered and certified to handle organic strawberries. If brokers or retail agents will take possession of the product when it leaves cooling or other facilities, they are required by law to keep adequate chain-of-custody records from shipment to sales point. Retailers must also keep records and must be registered with CDFA for such purposes.

At the time of this writing, organically produced strawberry transplants are not available, so you can use transplants from conventional nurseries to establish your organic strawberry field. You and the nursery need to follow all phytosanitary regulations for these transplants. All seeds and transplants used in organic strawberry production must not be treated with prohibited substances once established in the field, and this includes the seeds for any cover crops that you use. If you make a reasonable attempt to locate organic cover crop seed but are not able to find any, you may then use nonorganic (but still untreated) cover crop seed. Unlike strawberry transplants, many organic cover crop seeds are now commercially available.

REFERENCES/RESOURCES

Dimitri, C., and C. Greene. 2002. Recent growth patterns in the U.S. organic foods market. Washington, D.C.: USDA Economic Research Service Publication AIB-777. Accessed at www.ers.usda.gov/Publications/aib777.

Klonsky, K. 2010. A look at California's organic agriculture production. Davis: Giannini Foundation of Agricultural Economics, University of California, Davis. Accessed at http://agecon.ucdavis.edu/extension/update/articles/v14n2_3.pdf.

Klonsky, K., and K. Richter. 2005. A statistical picture of California's organic agriculture 1998–2003. Davis: University of California Agricultural Issues Center. Accessed at http://aic.ucdavis.edu/research1/organic.html.

Klonsky, K., and L. Tourte. 1998. Statistical review of California's organic agriculture 1992–1995. Davis: University of California Agricultural Issues Center Statistical Brief No. 6 (May). Accessed at http://aic.ucdavis.edu/publications/oldanrpubs/organicstatistical.pdf.

Klonsky, K., L. Tourte, R. Kozloff, and B. Shouse. 2002. Statistical review of California's organic agriculture 1995–1998. Oakland: University of California Agriculture and Natural Resources. Publication 3425. Accessed at http://aic.ucdavis.edu/research/misc/Organic1995-98.pdf.

OTA (Organic Trade Association). 2010. U.S. organic industry overview. Accessed at www.ota.com/pic/documents/2010OrganicIndustrySurveySummary.pdf.

USDA ERS (USDA Economic Research Service). 2011. Organic agriculture: Organic market review. Accessed at www.ers.usda.gov/briefing/organic/demand.htm.

Internet Resources

ATTRA (National Sustainable Agriculture Information Service), http://attra.ncat.org/organic.html.

California Organic Program (Processors), www.cdph.ca.gov/programs/Pages/FDB.aspx.

California Organic Program (Producers and processors), www.cdfa.ca.gov/is/i_&_c/organic.html.

eOrganic, www.eorganic.info.

Organic Farming Compliance Online Handbook, www.sarep.ucdavis.edu/organic/complianceguide/.

Organic Materials Review Institute, www.omri.org.

Scientific Congress on Organic Agricultural Research (SCOAR), www.ofrf.org/networks/scoar.html.

USDA National Organic Program, www.ams.usda.gov/nop.

Washington State Department of Agriculture Organic Food Program, http://agr.wa.gov/Portals/Org/.

SUSTAINABILITY INDICATORS

PHOTO: KATHY KEATLEY GARVEY

13

Sustainability Indicators in Organic Strawberries: Making the Conversion to Organic—and Beyond

STEPHEN R. GLIESSMAN

Strawberry growers in California have a reputation for being innovators and experimenters, willing to adopt new practices when they perceive some potential for benefit. This has been especially true in organic strawberry production, where over the past 20 years creative farmers have made changes in farming techniques that challenge conventional approaches to agricultural practices. Large increases in acreage devoted to the organic production of many crops, including vegetables and strawberries, have been apparent for more than a decade (USDA ERS 2005). In California alone, average annual organic sales grew by 15 percent while acreage increased by an estimated 10 percent per year between 1992 and 2003 (Klonsky and Richter 2005). A recent report from the USDA Economic Research Service (USDA ERS 2009) describes how, even with this growth, the demand for organic products continues to exceed the supply.

As growers continue to transition to organic production, we also continue to face the question of how sustainable these new organic agricultural systems really are. When we examine farming systems as ecological systems, more broadly known as agroecosystems, and when we base their design and management on the science of agroecology, we begin to realize that farmers and researchers need to work together to ensure that these new organic agroecosystems do not simply replace one set of problems with another. Defined as the application of ecological concepts and principles to the design and management of sustainable food systems (Gliessman 2007), agroecology offers a set of guiding principles for sustainability in the conversion to organic production.

"Sustainability" means that ecological soundness, economic viability and equity, and social well-being are valued simultaneously in the agroecosystem (Gliessman 2001). This definition implies that it is not enough merely to substitute conventional inputs and practices with organically acceptable alternatives. It is not enough to use an approach that is determined primarily by market demands but does not take into account the economic and social health of the grower, farmworkers, and agricultural communities in which food is produced. We need to develop a much broader set of tools to evaluate the sustainability of the conversion process. The new focus must be on the entire production and distribution system, from the field to the table, as strawberry growers link more directly with consumers in the development of sustainable food systems.

PRINCIPLES GUIDING THE CONVERSION PROCESS

The process of conversion to organic production can be complex, requiring changes in field practices, day-to-day management of the farming operation, planning, marketing, and even overall philosophy. The following can serve as general guidelines for the process of transition and the research that is necessary to make needed changes (Gliessman 2007):

- Shift from through-flow nutrient management to a recycling of nutrients, with increased dependence on natural processes such as biological nitrogen fixation, cover crops, composts, and mycorrhizal relationships, rather than primary reliance on purchased nutrient inputs. This will require research that determines whether nitrogen losses in leaching, runoff, crop removal, and volatilization can be matched by biological nitrogen fixation to meet total nitrogen needs and to achieve synchrony between nitrogen availability and nitrogen need for organic strawberry plants in California.

- Use renewable sources of energy, such as solar or wind power for pumping water or biodiesel fuel for tractor operation, rather than conventional, nonrenewable energy sources. Documentation is needed from farms that have installed alternative energy systems, such as Lundberg Farms in Northern California.
- Reduce or eliminate the use of nonrenewable off-farm inputs that have the potential to harm the environment or the health of farmers, farmworkers, or consumers. For example, release natural enemies instead of spraying synthetic biocides, or plant hedgerows that serve as pest trap crops so you can concentrate control practices on the noncrop part of the field. There are many successful examples of the release of beneficial arthropods in strawberry crops—including the release of the predator mite *Phytoseiulus persimilis* for twospotted spider mite control—but more research is needed to convincingly demonstrate the effectiveness of hedgerows.
- When materials must be added to the system, use naturally occurring, nonsynthetic materials instead of synthetic, manufactured inputs whenever possible. Such materials still need to be tested for efficacy and economic viability.
- Design the farm so as to "manage" pests, diseases, and weeds preventively rather than "control" them with remedial actions. Effective management strategies need to be designed and tested for all growing regions in California.
- Encourage the biological relationships that can occur naturally on the farm, instead of reducing or simplifying them. As the transition to biologically diverse strawberry systems progresses, considerable field research will be needed to study effective interactions.
- Make appropriate matches between cropping patterns and the productive potential and physical limitations of the farm landscape to reduce the need to modify the environment. Each growing region in California will need its own cropping patterns, uniquely suited to its local environment.
- Use a strategy that adapts the biological and genetic potential of agricultural plant and animal species to the farm's natural ecological conditions, rather than modifying the farm to meet the needs of the crops and animals. As new crop varieties are developed for organic systems, new materials will need to be tested in each strawberry growing region.
- Emphasize the conservation of soil, water, energy, and biological resources. All elements of the production system must be sustainable.

The integration of these principles incorporates the idea of long-term sustainability into overall agroecosystem design and management and creates a synergism of interactions and relationships on the farm that will eventually lead to the development of sustainable agroecosystems. As ecologically based alternatives are developed, we need to test them for their contribution to the overall economic viability of strawberry production. The emphasis on particular principles in the transition to organic strawberry production will of course be specific to the crop, but all of what we learn can contribute greatly to the overall conversion process. By measuring and monitoring every indicator of sustainability, we will be able to get a sense of the long-term health and viability of the new farming system.

For many farmers, rapid conversion to organic strawberry production is neither possible nor practical. Regulations require a 3-year transition period, but 3 years may not be enough time to re-establish many ecological processes and relationships. As a result, many conversion efforts will proceed in slower steps toward the ultimate goal of sustainability, meanwhile making some changes to meet organic standards. Studies of the conversion process are still relatively limited (for examples see Swezey et al. 1994, 1999; Hendricks 1995; Gliessman et al. 1996; Gliessman and Rosemeyer 2010). They tell us that much research still needs to be done to improve yields and pest management results as well as to improve indicators of sustainability.

Current research efforts point out three distinct levels of conversion. These help us describe the steps that farmers actually take in converting from conventional agroecosystems, and they can also serve as a map outlining a step-by-step, evolutionary conversion process that organic systems can follow in order to achieve sustainability. In addition, they help to categorize ways in which agricultural research has been applied to strawberries.

Level 1. Increase the efficiency of conventional practices in order to reduce their use and the consumption of costly, scarce, or environmentally damaging inputs.

This step is what we might call "pre-organic." It is also synonymous with responsible, progressive conventional growing or best management practices (BMPs). The goal is to use conventional inputs more efficiently so the farm will require fewer inputs and the negative impacts of their use will be reduced. This approach has been the primary emphasis of much of the conventional agricultural research in strawberries and has been the source of numerous innovative agricultural technologies and practices. Examples include optimal strawberry plant spacing and density, improved machinery for soil fumigation, pest monitoring for improved pesticide application, improved timing of operations, and precision farming for optimal fertilizer and water placement. Although these kinds of efforts reduce the negative impacts of conventional agriculture, they do not help break its dependence on external human inputs and do not make for a production system that can be certified as organic.

Level 2. Replace conventional inputs and practices with organic practices.

We might call this phase "commercial organic." The goals at this level of conversion are to replace resource-intensive and environment-degrading products and practices with those that are more environmentally benign, and more importantly to meet the standards required by certified organic markets. Conventional inputs are replaced with those that are deemed acceptable by organic standards, especially any inputs that are synthetic in origin. Most organic farming research for strawberries has emphasized this kind of approach. Examples of alternative practices include the use of nitrogen-fixing cover crops or nitrogen-rich, organic, natural products to replace synthetic nitrogen fertilizers, the use of biological control agents rather than synthetic pesticides, and the shift to organic soil amendments. At this level, the basic agroecosystem structure is not greatly altered (i.e., strawberry monoculture is still common), so it is subject to many of the same problems as occur in conventional systems.

Level 3. Redesign the agroecosystem so that it functions on the basis of a new set of ecological processes.

We might call this level "sustainable organic." At this level, overall system design strives to eliminate the systemic causes of many of the problems that still exist at levels 1 and 2. Rather than working to find more efficient ways to solve problems, this method prevents problems from arising in the first place. Whole-system conversion studies encourage the understanding of yield-limiting factors in the context of agroecosystemic structure and function. Problems are recognized, and thereby prevented, by means of internal site- and time-specific design and management approaches rather than by the application of external inputs. An example is the diversification of farm structure and management through the following techniques:

- Rather than attempting to grow strawberries every year in the same field, the grower rotates strawberries with alternative crops that are specifically selected for their ability to suppress populations of soilborne diseases while simultaneously providing an alternative source of income for the farmer. The best-known example of this focus is the work of Muramoto and associates in the Elkhorn Slough watershed of Monterey County, where a 5-year experiment demonstrated that rotations involving crops such as broccoli and spinach, combined with cover crops that included wild mustard, could shorten the time period for replanting strawberries to 3 years while maintaining economically profitable returns in an organic system (Muramoto et al. 2006).
- Intercropping strawberries with other crops or even with noncrop plants that can function as trap crops for insect pests and attract and retain beneficial insect predators and parasitoids. The pioneering research of Swezey and Pickett and their associates (Swezey, Nieto, and Bryer 2007; Pickett et al. 2009) has shown that the insertion of a row of alfalfa in place of every fortieth or fiftieth row of strawberries can effectively control the western tarnished plant bug *(Lygus hesperus)* and retain beneficial arthropods while maintaining the field's economic profitability.

- Integrate natural areas, such as hedgerows or restored natural vegetation, into the farm landscape to provide a broader array of environmental services beyond crop yields—services such as erosion control, conservation of biodiversity, or carbon sequestration. Research trials with hedgerow plantings in strawberry growing areas of Santa Cruz County have provided valuable information on how hedgerows have the potential to harbor beneficial insects and enhance biological control (Pisani-Gareau, Shennan, and Letourneau 2007).

In terms of research, agronomists and other agricultural researchers have done a good job on the transition from level 1 to level 2, but research on the transition to level 3 has only just begun. The field experiments with rotations, rows of alfalfa, and the introduction of hedgerows cited above are examples of such research. It is important to note that all level-3 research must be accompanied by detailed economic analysis to ensure that any changes made in the design of the production system also result in economic benefits to the grower. An agroecological focus on the conversion process provides the basis for this type of research. In time, it will help us find the answers to larger, more difficult questions such as what sustainability is and how we will know we have achieved it.

ON-FARM CONVERSIONS

As strawberry farmers undertake to convert their farms to organic management, it becomes clear that there is a need to develop systems for evaluating and documenting the success of these efforts and the changes they engender in the functioning of the agroecosystem. The amount of time it takes to complete the conversion process depends on the type of crop or crops being farmed, ecological conditions at the farm location, and the farm's history of management and input use. For short-term annual crops, the time frame for conversion might be as short as 3 years; for perennial crops, it is probably 5 years or longer. For intensively managed strawberries, figure on at least 5 years, especially when you consider the need to replace conventional soil fumigation with alternative crop rotations.

Study of the conversion process involves data collection and analysis on several levels. You need to

- examine the changes in ecological factors and processes over time as observed through monitoring and sampling, nutrient cycling rates, organic matter production, soil biodiversity, and other important ecological characteristics of the farm system
- observe how strawberry fruit yields change with changes in practices, inputs, designs, and management
- understand the changes in energy use and labor that accompany the previously noted changes
- evaluate the economic costs and benefits of all changes and adjustments made to the design of the strawberry production system
- based on accumulated observations and research, identify key indicators of sustainability that are "farmer-friendly" and can be adapted to on-farm, farmer-based monitoring programs, and are also grounded in an understanding of ecological sustainability (examples include such factors as soil quality and health, water availability and quality, pest-predator ratios and balance, and nutrient budgets)

You can re-examine research results, site-specific ecological factors, farmer skill and knowledge, and new techniques and practices each season to determine whether you need to make any modifications to overcome identified yield-limiting factors. Ecological components of the system's sustainability become identifiable at this time, and eventually these can be combined with other factors for an analysis of economic and social sustainability.

SUSTAINABILITY INDICATORS

The ecological framework we have described here provides a set of ecological parameters that you can study and monitor over time in your strawberry production system to assess its movement toward or away from sustainability. For each parameter, agroecological knowledge suggests a general condition or quality necessary for the sustainable functioning of the system—for example, high biodiversity, increased organic matter content, and improved

Table 13.1. Parameters related to strawberry agroecosystem sustainability, as discussed at the Salinas Organic Strawberry Production Short Course, February 2003

A. Plant Indicators

General plant health and development. What are the indicators of a healthy plant?

a. root volume ("grow roots in order to grow fruit"), based on research on the relationship between root volume and fruit yield
b. root/shoot ratio and plant development
c. leaf surface area, based on research on the relationship between leaf area and fruit yield
d. canopy diameter
e. fruit production (at least 3,750 trays per acre); investigate economic viability of yield levels
f. proportion of fresh market yield to processing to culls
g. availability of varieties appropriate for organic systems, based on breeding programs that select for organic production conditions

B. Soil Factors

Soil conditions over the long term and the short term.

a. nutrient uptake efficiency (inputs − crop outputs = 0; if greater than 0 and the difference is not found in the soil, then it was a loss from the system)
b. organic matter content in the topsoil and its quality
c. proper nutrition and nutrient availability (total N at least 2.8% in leaflet dry weight; at least 500 ppm NO_3 in petiole dry weight)
d. mineralization rates and timing to ensure nutrient availability when needed, especially for nitrogen
e. improvement in soil physical properties (compaction, drainage, etc.)
f. amount of compost added in relation to crop needs and soil conditions
g. weed population density and composition (proportions of "bad" to "good" weeds; for example, weeds that harbor *Verticillium* vs. weeds that fix nitrogen)

C. Water Quality Indicators

On-farm water use efficiency and water quality issues:

a. pollution levels of surface water and groundwater sources
b. soil erosion rates and control measures
c. amount of nitrates in groundwater (must meet EPA standards)
d. amount of water used per acre
e. degree of salinization of soils and groundwater

D. Plant Pathogens

On the plant and in the soil:

a. total microbial biomass in the soil (with a focus on beneficial organisms)
b. threshold levels of key soilborne plant pathogens (1.0 *microsclerotia*/gram soil for *Verticillium*)
c. species of *Pythium* present in the soil (watch out for *P. ultimum*)
d. crown rot complex dynamics
e. detrimental vs. beneficial rhizosphere colonizers
f. canopy coverage or diameter in relation to specific rhizosphere populations
g. wilt severity due to the degree of *Verticillium* infestation
h. microbial competition potential-diversity indices
i. incidence of fruit and foliar diseases over time
j. susceptibility vs. resistance

E. Arthropod Population Dynamics

Ecology and management of arthropods:

a. species and numbers of spiders as predators
b. population thresholds for key pests, based on monitoring (e.g., 5 bugs/50 plants for lygus)
c. population levels of natural enemy species (predators and parasitoids)
d. effectiveness of degree models for prediction and management
e. arthropod populations on trap crops
f. amount of arthropod damage to the crop

F. Ecosystem-Level Characteristics

Emergent qualities of the entire agroecosystem as new components are integrated:

a. crop system diversity (number of crops, level of farmscaping, number of rotations, integration of natural vegetation)
b. efficiency of nutrient budgets and nutrient cycles
c. level of integration of animals in farming systems (especially as sources of soil amendments)
d. weed seed bank composition over time
e. degree of interaction and integration of different agroecosystem components
f. system-level management
g. sources of energy and efficiency of use

G. Economic Indicators

Farm viability and profitability, both on-farm as well as in the surrounding farming communities:

a. farm gate value of crop and total value of sales
b. adequate (fair, equitable, and sufficient) wholesale price for the farmer
c. net return above total costs
d. opportunities for new markets
e. percentage domestic vs. imported market share
f. overall farm profitability
g. cost of registration, certification, and inspections
h. product quality and consistency

H. Institutional Indicators

Policy, education, research, and outreach to organic strawberry farmers:

a. availability of funding for research on sustainable strawberry systems
b. education programs in organic agriculture
c. training programs for Extension
d. local government support

I. Social Indicators

Cultural and social health of agricultural communities:

a. number of acres in production
b. number and age of growers
c. diversity of growers
d. number of acres per farm
e. turnover of farmers
f. grower changes in philosophy
g. consumer changes in philosophy

J. The Red Edge

The California Strawberry Commission's focus on the future of organic strawberries to increase consumption:

a. links to quality of human aging
b. improved heart protein levels
c. cancer protection and reduction in incidence
d. eat six strawberries a day

K. The Sustainable Organic Red Edge

Linking ecological, economic, and social components in organic strawberry agroecosystems:

a. superior taste and flavor of organic berries
b. more fair trade and direct-market systems
c. development of redesigned and diversified agroecosystems
d. environmental protection
e. community support for organic farmers
f. informed consumer choices
g. sustainable in all aspects (ecological, economic, and social)

topsoil. The specific rate, level, value, or status of each of these parameters that, taken together, indicate a condition of sustainability, however, will vary from farm to farm because of differences in farm type, resources used, local climatic conditions, and other site-specific variables. Each system, therefore, must be studied on its own to generate a set of system-specific indicators of sustainability.

The parameters listed in table 13.1 provide a framework for research concerning a sustainable agroecosystem for organic strawberries. The table presents data for some parameters, where available, but for most parameters the researchers and growers participating in the 2003 Salinas Organic Strawberry Production Short Course simply pointed out areas where research and actual values are needed.

CONCLUSIONS

The remarkable and sustained growth of the market for organic strawberries has led to a corresponding overall increase in production and the acreage devoted to this crop (CDFA 2008; Klonsky and Richter 2007). The research establishment has responded to this growth by helping growers develop a remarkable array of inputs and practices that meet the norms required for organic certification. But when we apply the norms of sustainability, we must engage research at another level. Simple substitution of organically acceptable inputs for conventional ones is not by itself sufficient. Strawberry agroecosystems must be redesigned to be ecologically resistant to the problems that those conventional inputs have been used to solve, and must at the same time maintain profitability for the grower. Researchers must employ the system-level understanding of agroecology and consider the long-term interplay of the ecological, economic, and social components of sustainable farming. By developing the indicators we have discussed in this chapter, growers should be able to look beyond the norms of organic certification to the broader criteria of sustainability while still maintaining economic viability.

REFERENCES/RESOURCES

CDFA (California Department of Food and Agriculture). 2008. Organic sales reports: Gross yearly sales, acreage by commodity for 2007. Accessed at www.cdfa.ca.gov/is/docs/COMMYearlysalesACREAGE2007Producers.pdf.

Gliessman, S. R. 2001. Agroecosystem sustainability: Towards practical strategies. Advances in Agroecology Series. Boca Raton: CRC Press.

———. 2007. Agroecology: The ecology of sustainable food systems, Second edition. Boca Raton: CRC/Lewis Publishers.

Gliessman, S. R., and M. E. Rosemeyer (eds). 2010. Converting to sustainable agroecosystems: Principles, processes, and practices. Advances in Agroecology Series. Boca Raton: CRC/Taylor & Francis Publishers.

Gliessman, S. R., M. R. Werner, S. L. Swezey, E. Caswell, J. Cochran, and F. Rosado-May. 1996. Conversion to organic strawberry management changes ecological processes. California Agriculture Journal 50:24–31.

Hendricks, L. C. 1995. Almond growers reduce pesticide use in Merced County field trials. California Agriculture Journal 49:5–10.

Klonsky, K., and K. Richter. 2005. A statistical picture of California's organic agriculture 1998–2003. Davis: University of California Agricultural Issues Center. Accessed at http://aic.ucdavis.edu/research1/organic.html.

———. 2007. Statistical review of California's organic agriculture 2000–2005. Davis: University of California Agricultural Issues Center. Accessed at http://aic.ucdavis.edu/publications/Statistical_Review_00-05.pdf.

Muramoto, J., S. R. Gliessman, S. T. Koike, C. Shennan, D. Schmida, R. Stephens, and S. L. Swezey. 2006. Maintaining agroecosystem health in an organic strawberry/vegetable rotation system. Part 5: The final result. Annual Meeting Abstracts. Annual Meeting of the American Society of Agronomy, Indianapolis, Indiana, Nov. 12–16, 2006. Annual meetings abstract CD.

Pickett, C. H., S. L. Swezey, D. J. Nieto, J. A. Bryer, M. Erlandson, H. Goulet, and M. D. Schwartz. 2009. Colonization and establishment of *Peristenus relictus* (Hymenoptera: Braconidae) for control of *Lygus* spp. (Hemiptera: Miridae) in strawberries on the California central coast. Biological Control 49:27–37.

Pisani-Gareau, T., C. Shennan, and D. Letourneau. 2007. Hedgerows: Enhancing conservation biological control potential in California annual vegetable systems. (Abstract). 92nd Annual Meeting Ecological Society of America, San Jose, California, Aug. 2007.

Swezey, S. L., P. Goldman, R. Jergens, and R. Vargas. 1999. Preliminary studies show yield and quality potential of organic cotton. California Agriculture Journal 53:9–16.

Swezey, S. L., D. Nieto, and J. Bryer. 2007. Control of western tarnished plant bug, *Lygus hesperus*, Knight (Hemiptera: Miridae) in California organic strawberries using alfalfa trap crops and tractor-mounted vacuums. *Environmental Entomology* 36:1457–1465.

Swezey, S. L., J. Rider, M. Werner, M. Buchanan, J. Allison, and S. Gliessman. 1994. Granny Smith conversions to organic show early success. California Agriculture Journal 48:36–44.

USDA ERS (USDA Economic Research Service). 2005. Organic production: The economics of food, farming, natural resources, and rural America. USDA Economic Research Service. Accessed at www.ers.usda.gov/data/organic/.

———. 2009. Emerging issues in the U.S. organic industry. Economic Information Bulletin No. 55:1–28.

Index

Note: Page numbers in **bold type** indicate major discussions. *Italic type* is used to indicate tables, e.g., *19t,* and figures, e.g., *61f.*